An Introduction to Biological Rhythms

TERTIARY LEVEL BIOLOGY

A series covering selected areas of biology at advanced undergraduate level. While designed specifically for course options at this level within Universities and Polytechnics, the series will be of great value to specialists and research workers in other fields who require a knowledge of the essentials of a subject

Other titles in the series:

TERTIARY LEVEL BIOLOGY

An Introduction to Biological Rhythms

David S. Saunders, B.Sc., Ph.D.

Reader in Zoology
University of Edinburgh

A Halsted Press Book

John Wiley and Sons

New York

Blackie & Son Limited
Bishopbriggs
Glasgow G64 2NZ

450 Edgware Road
London W2 1EG

Published in the U.S.A. by
Halsted Press,
a Division of John Wiley and Sons Inc.,
New York

® 1977 D. S. Saunders
First published 1977

Library of Congress Cataloging in Publication Data
Saunders, David Stanley, 1935-
An introduction to biological rhythms

(Tertiary level biology)
"A Halsted Press book"
Bibliography: p.
Includes index.

1. Biological rhythms. I. Title.
QH527. S24   1977   574.1   76-48623
ISBN 0-470-99019-8

Phototypesetting by
Print Origination, Merseyside, Lancs

Printed in Great Britain by
Thomson Litho Ltd., East Kilbride, Scotland

Preface

CIRCADIAN AND OTHER BIOLOGICAL RHYTHMS WHOSE PERIODS ARE meaningful in terms of "real" (i.e. planetary) time, are to be observed at all levels of organization, in activities varying from respiration, photosynthesis and cell division in unicellular (eukaryotic) organisms to the sleep-wake patterns in birds and mammals, including man. A study of these biological "clocks" may thus become a "potted" version of eukaryotic biology and one which, moreover, serves as a useful link subject between ecology and behaviour on the one hand, and physiology and biochemistry on the other. The study of biological rhythms is also a truly integrated biology; a student of animal circadian systems or of insect photoperiodism, for example, would soon feel at home in a botany department interested in rhythms and clocks. Indeed, some of the leaders in the field of biochronometry, such as Erwin Bünning and Colin Pittendrigh, are personally involved in research at all levels of organization, and with plants as well as animals.

This book is aimed at advanced undergraduates (i.e. third-year or, in Scottish Universities, fourth-year undergraduates), and at those beginning a research career in biochronometry. In most universities, third-year students are required to read original research papers, and final-year students frequently spend at least part of their time in research projects. This book, therefore, has been specifically designed to cover the field of biochronometry at a level intermediate between that of an elementary text and a research review; it should introduce advanced students to the subject more gently than by presenting them with a list of references to journal articles. The bibliography is therefore restricted, but includes recent or comprehensive reviews as well as pioneering papers. Much of the research activity in biological rhythms has been carried out in Germany and, for this reason, references to work in the German language cannot be omitted; these, however, have been provided with an English title. Some of the terms and abbreviations may be new to students, particularly in universities where courses in biochronometry have not been a strong feature; but the

v

terms used have gained wide international acceptance, and a glossary is provided for the help of the reader. Finally, as is consistent with an "integrated" biological subject, examples are quoted from both the animal and the plant world—although the former predominate, reflecting my own particular interest in animals, especially the insects. Reference to botanical examples occurs mainly where plants show a particular property better than animals, or where the first observations were made with plant material.

DAVID S. SAUNDERS
Department of Zoology
University of Edinburgh

Contents

vii

viii

CHAPTER ONE

PERIODICITY IN THE ENVIRONMENT AND IN THE ORGANISM

Periodicity in the environment

Life on Earth is and always has been exposed to strong and rhythmic environmental changes caused by planetary movements. Thus the rotation of the Earth on its polar axis gives rise to the dominant cycle of day and night; the revolution of the Earth around the Sun gives rise to the unfailing procession of the seasons; and the more complicated movements of the Moon in relation to the Earth and the Sun give rise to the lunar month and to the tidal cycles. Since the endogenous biological rhythms which form the subject matter of this book have evolved in response to these environmental periodicities, and are now entrained by them, we will examine these planetary movements in greater detail.

The Earth spins on its axis once a day, so that the Sun appears to rise in the east and sink in the west. Daylength is, of course, the same at any one latitude, but local time varies with the longitude, the sun rising or setting about one hour later for every 15° of longitude as one travels west. The time taken for the Earth's rotation, measured with reference to the stars (which for all practical purposes are infinitely remote) is called the *sidereal day,* and is 23 hours 56 minutes 4 seconds. For an observer or stationary organism on this planet, however, the *solar day,* or the time interval between two successive noons, is clearly of more significance. The solar day is about 24 hours 4 minutes, but is not constant because the Earth's orbit is somewhat elliptical. According to Kepler's Second Law, a planet moves faster when nearer to the Sun (at perihelion) than when further from the Sun (at aphelion); consequently the solar day varies slightly during the year, but never more than 16 minutes* from the mean (24 hours 3 minutes 57 seconds). It is clear, however, that although the solar day is variable and frequently different from 24 hours, 24 hours is a reasonable approximation

*The solar day is the same as the mean solar day four times in the year (April 15, June 14, September 1, December 25). The maximum deviation (16 minutes shorter) occurs on about November 3. This deviation from the mean is called the Equation of Time.

1

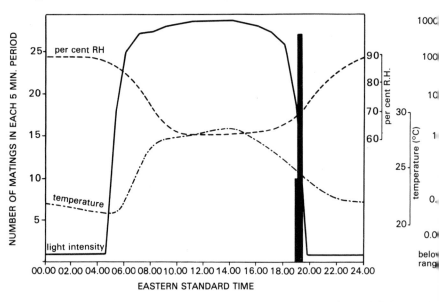

Figure 1.1 Diurnal variation in light intensity, temperature and relative humidity in natural conditions. The histogram shows the number of matings of the fruit fly *Dacus tryoni* in each 5-minute period. Note that mating takes place only at dusk. After Tychsen, P.H. and Fletcher, B.S. (1971), *J. Insect Physiol.,* **17**, 2139–2156, Fig. 1, Pergamon Press, Oxford.

to the natural day-night cycle to which an organism living on the surface of this planet is exposed.

The cycle of day and night is important for a number of obvious reasons. Practically all of the energy entering the biotic environment is derived from the Sun, being incorporated by green plants during the day but not at night. Correlated with this rhythm in light intensity are also rhythms of temperature and humidity (figure 1.1); in other words, the three most important environmental variables have a strong 24-hour periodicity.

The year can be measured in a number of ways. Perhaps the most important in the present context is the *sidereal year,* measured with reference to the stars, and representing the true revolutionary period of the Earth around the Sun (365 days 6 hours 9 minutes 10 seconds). As noted above, the Earth is not always at the same distance from the Sun. The seasons, however, are not caused by this factor, but occur because the axial inclination of the Earth to the "perpendicular" (or the angle between the ecliptic and the "celestial equator") is about 23½° and, as the Earth revolves around the sun, the face presented to the source of radiation varies (figure 1.2*a*). When viewed from the Earth, the Sun appears at its northernmost point in the sky (declination 23½°N) around June 22 to give the summer solstice in the Northern Hemisphere and the winter solstice in the Southern Hemisphere. Conversely, the Sun is at its southernmost point (23½°S) around December 22 to give the summer solstice in the Southern Hemisphere and the winter solstice in the Northern. Twice a year the sun crosses the "celestial equator", once when moving from south to north (about March 21), and once when moving from north to south (about September 22), giving rise to the vernal and autumnal equinoxes, respectively.

The most important consequence of this progression is that the number of hours of light per day varies with the season. At the equinoxes, all parts of the globe experience a 12-hour day and a 12-hour night. After the autumnal equinox, however, days shorten to the winter solstice and then lengthen; and after the vernal equinox the days lengthen to the summer solstice and then shorten. The absolute length of the day at any one time is, of course, a function of latitude (figure 1.2b). It is also obvious that the

(a)

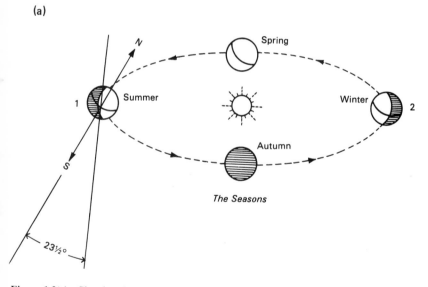

Figure 1.2(a) Showing the angle of inclination (23½°) of the Earth's axis, and the four seasons.

(b) The seasonal changes in daylength at different latitudes in the northern hemisphere. After Danilevskii.

(c) Seasonal changes in daylength and periods of civil and nautical twilight in a locality 55° North of the equator. A = length of solar day. B= solar day plus civil twilight. C = solar day plus nautical twilight.

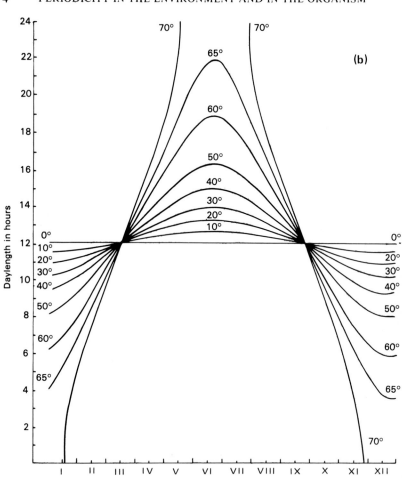

axial inclination of the Earth, and the seasons it produces, results in the familiar cycles of temperature—winter temperatures being lower than those in the summer because of the shorter day and the smaller angle of incidence of the solar radiation.

Daylength is of particular importance to the subject matter of this book because it has been adopted by a wide variety of animals and plants as a reliable and "noise-free" indicator of season. However, if daylength or nightlength is to be "measured" by an organism, presumably with the aid of a photoreceptor which responds to a threshold in light intensity, the

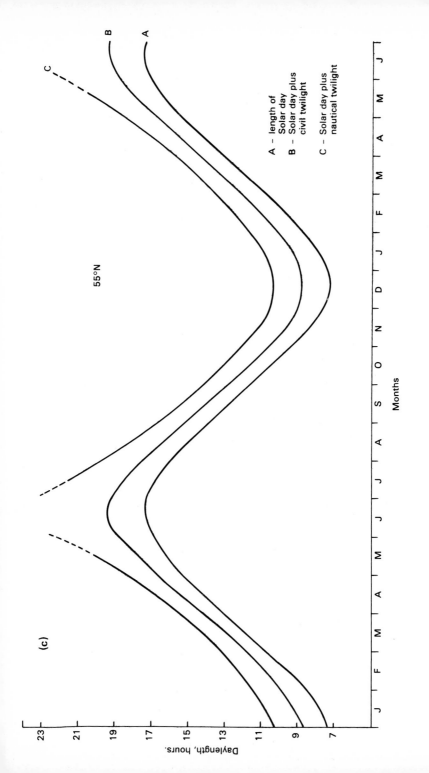

55°N

A – length of
    Solar day
B – Solar day plus
    civil twilight
C – Solar day plus
    nautical twilight

A
B
C

Daylength, hours.

Months

(c)

periods of *twilight* are also of significance. The shortness of the twilight period close to the equator is well known; so also is the much longer and variable period of twilight at higher latitudes. Daylength, including the two periods of *civil twilight*, is defined as the time required for the upper limb of the Sun to traverse an arc from 6° below the horizon in the east to a point 6° below the horizon in the west. *Nautical twilight* and *astronomical twilight* are based on similar calculations, but involve angles of 12° and 18° respectively. Some of these changes in daylength and in the twilight periods are shown in figure 1.2c for a locality 55° north of the equator. The *effective* threshold to which a particular animal or plant may respond, depends on the threshold for the photoreceptor involved, but it frequently lies between civil twilight and nautical twilight. At high latitudes the twilight periods during the summer become very long, and the night is never really dark (above the Arctic circle the sun never sinks below the horizon at midsummer). Nevertheless, except at the very highest latitudes, there is always a sufficient difference between night and day to constitute a 24-hour rhythm of light intensity. In the continuous light of the polar day there is usually a diurnal variation in temperature.

   The combined revolutions of the Moon and the Earth, and of both of them around the Sun, give rise to the more complicated periodicities of moonlight and tide, both of which are important variables in the lives of animals and plants—particularly those which live in the sea. The Moon completes its orbit around the Earth in a month: as with "days" and "years", however, there are several ways to define a "month", two of which are important in the present context. The *sidereal month,* measured with reference to the stars and representing the true lunar orbit, is 27.32 days. The synodical month (or *lunation*), on the other hand, is the interval between successive new moons or full moons as seen from a point on the Earth's surface, and occupies 29.53 days. Viewed from the Earth, the Moon passes through successive phases, from new moon to crescent, half, gibbous and full (figure 1.3a), according to the relative positions of the Earth and Moon in relation to the Sun. The intensity of moonlight, therefore, varies with the cycle of lunation: at full moon it is rarely greater than about 0.7 lux, or 0.28 μW cm^{-2}. Many animals and plants are known to use the time cues provided by the lunar cycle to synchronize functions such as reproduction within the population. A combination of the Moon's revolution around the Earth (27.32 days) and the Earth's rotation on its

Figure 1.3(a)   The phases of the moon. (b) Spring and Neap tides. (c) Tidal cycles during one month in an area with a high tidal range.

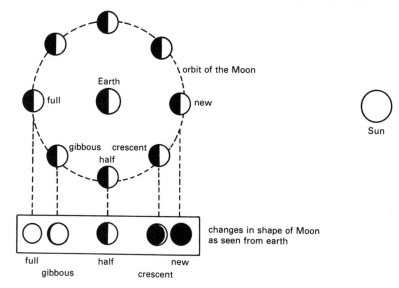

(a) *Phases of the Moon*

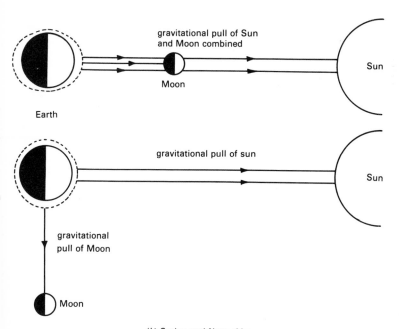

(b) *Spring and Neap tides*

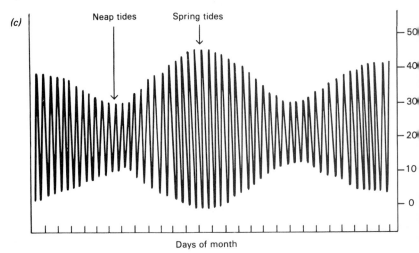

Probably the most important effect of the Moon's revolution around the Earth is the control of the complex *tidal cycles*. The gravitational pull of the Moon causes the waters of the Earth to "pile up" underneath it (and, of course, on the other side of the Earth). As the Earth spins on its axis, this "heap" sweeps around the world, giving two high tides in every lunar day (i.e. once every 12.4 hours)—although variations in ocean depth, ocean currents, and the irregular shapes of the land masses make this simple picture in reality rather more complex. On many coasts, for example, the two tidal crests per day are equal or nearly so; on other coasts the two tides are of markedly different height (the *semi-diurnal inequality*), or only one high tide may occur every 24.8 hours. The time intervals between the two tides may also vary, although the sum of the two intervals is always close to that of the lunar day (24.8 hours). The Sun also exerts a gravitational pull

axis (~24 hours) results in the Moon rising a little later every night, producing the so-called lunar-day (~24.8 hours). This *retardation* is normally more than 30 minutes every night, but may be as much as an hour. In the Northern Hemisphere, however, the full moon closest to the autumnal equinox undergoes a nightly retardation of no more than 15 minutes; hence the popular belief that this "harvest moon" rises at about the same time for several nights in succession. This romantic, and seemingly trivial observation, might be important in considering the Moon's potential interference with night-length measurement in photoperiodism (Chapter 5).

on the waters of the Earth, although of a smaller magnitude. Therefore, when the Sun and the Moon are pulling in the same direction, the tidal range is particularly high (*spring tides*), whereas when the Moon and the Sun are at an angle of 90° (i.e. the Moon is at *quadrature*) the tides are less high (*neap tides*) (figure 1.3*b*). The time interval between successive spring tides or successive neap tides is therefore half the lunar period, or 14.77 days. Figure 1.3*c* illustrates the tidal cycle in an area with a high tidal range.

Finally, since organisms have evolved in these strong environmental periodicities, it is relevant to point out that the periodicities themselves have changed over the aeons. Astronomers appear to agree that while the period of the Earth's revolution around the Sun has remained constant, the period of its rotation on its polar axis has slowed down, principally because of the dissipation of rotational energy by tidal forces. This deceleration is of the order of 2 seconds every 100,000 years, which means that the length of the day at the beginning of the Cambrian Period (600 million years ago) was only 21 hours, and that there were more than 420 such days in the Cambrian year. Palaeontological evidence for this increase in the number of days has been obtained from the study of fossil corals, which show a "circadian periodicity" in the deposition of their skeleton, probably consistent with the observation that modern corals take up calcium carbonate maximally during the day and minimally at night. Corals from the middle Devonian Period, for example, indicate that there were between 385 and 410 days in the Devonian year (Wells, 1963). Similar evidence has been used to calculate changes in the Moon's orbit: in the Precambrian era, 18.7 solar days per lunar month and a 9-hour tidal cycle are thus indicated (Turcote *et al.*, 1974). Of course, organisms have had plenty of time in their evolutionary history to respond to such changes, but these observations are a strong indication that the endogenous biological rhythms included in this book have been a feature of the organization of animals and plants throughout the period of organic evolution on this planet.

Periodicity in the organism: biological chronometry

The environmental periodicities outlined above find their counterparts in the biological rhythms which control many overt behavioural and physiological processes in plants and animals. During their long evolutionary history, organisms have developed various endogenous rhythmicities whose periods match those of the cycles of day and night (*circadian*, period~24 hours), lunation (*circalunar*, period~29 days), tide (*circatidal*, period~12.4 or 24.8 hours), the seasons (*circannual*, period~a year), or the time between successive spring low waters (semilunar or *circasyzygic*, period ~14.7 days). The best known of these rhythmic phenomena is the circadian (from *circa*, about; *dies*, a day); this will be examined in detail in the next few chapters. Rhythms with non-circadian periodicity, but with remarkably similar properties otherwise, will be discussed later in the book, together with other functions of the "circadian system" in which oscillatory processes are, or may be, used to measure time—as in Photoperiodism, the "time sense" of animals, and various forms of orientation behaviour. At the moment it is sufficient to note that circadian rhythms are those which *persist* when all environmental periodicities are excluded, and in this free-running condition show a *natural period* which is

(1) *close to* but rarely equal to *that of the solar day*

(2) surprisingly *accurate* (for a biological phenomenon)

(3) *temperature-compensated.*

In short, they possess the properties one would expect for a *biological clock* and, indeed, that would seem to be their modern biological function: they measure out the passage of time in much the same way as a pendulum.

Circadian rhythmicity may be observed at all levels of organization except, apparently, in the prokaryotes. *Unicellular organisms,* particularly algae, show circadian rhythms of phototaxis, cell division, photosynthesis and bioluminescence (Sweeney, 1969). *Fungi* such as *Neurospora* spp. and *Pilobolus* spp. show a 24-hour rhythm of sporulation, although other fungi may show a variety of rhythmic patterns with a non-circadian periodicity (Jerebzoff, 1965).

"Higher" *plants and animals* show a vast array of periodicities, some of the most frequently studied being leaf movements of plants, and the sleep-wake or activity cycles of insects, birds and mammals. At the *population* level, certain "once-in-a-lifetime" events, such as egg hatch, moulting, pupation, and adult emergence of insects, show a pronounced circadian rhythmicity in a mixed-age population. One of the most intensively investigated of such systems is the rhythm of pupal eclosion in the fruit fly

---

Figure 1.4   Nocturnal and diurnal activity.
(*a*) Onsets of running activity in the flying squirrel *Glaucomys volans* in natural conditions. The dotted line indicates the time of local sunset. After De Coursey, P. (1960), *Cold Spring Harbor Symp. quant. Biol.,* **25**, 49, Fig. 1.
(*b*) Onset and end of flying activity in the jackdaw in natural conditions. Solid line: sunrise and sunset; broken line: the sun at 6° below the horizon (civil twilight). After Aschoff and von Holst.
(*c*) Phototactic rhythm in *Euglena gracilis* in an artificial light cycle (*LD 12*:12) and then in continuous darkness (DD). After Pohl.

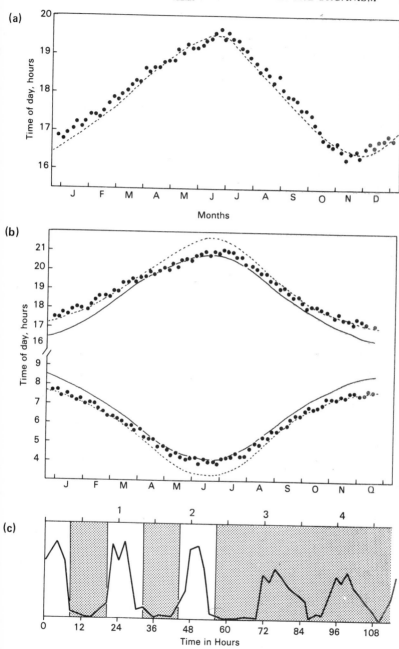

Drosophila pseudoobscura (Pittendrigh, 1954, 1960, 1966). Even *cultured cells, tissues and organs* may show persistent circadian rhythms when isolated from the whole organism. Thus in plants, isolated sections of leaves continue to show diurnal fluctuations in turgor pressure, growth and CO_2 output (Wilkins, 1960). In animals, isolated sections of the gut of hamsters show circadian rhythms of peristaltic activity (Bünning, 1973), isolated eyes of the marine mollusc *Aplysia californica* a rhythm of optic nerve activity (Jacklet, 1969), and isolated salivary glands from *Drosophila melanogaster* a rhythm in nuclear volume (Rensing, 1969).

What is the selective advantage of these biological rhythms?

Since one of the cornerstones of Darwinian evolution is Natural Selection, it is relevant to ask this question—and also, perhaps, to presume that biological rhythms have some adaptive value. The answer seems to lie in "temporal organization".

In a natural 24-hour periodicity of light and temperature, the circadian system of an animal or plant, which is also oscillatory and with a near-24-hour periodicity of its own, becomes *entrained* to that of the environment in much the same way that two physical oscillators will achieve mutual entrainment. Thus the immediate function of the daily changes in light intensity and, to a lesser extent, temperature, is that of an entraining agent, "time cue", synchronizer, or *Zeitgeber*: the last term, derived from the German literature, is the most commonly used and will be adopted here. When exposed to such an environmental Zeitgeber, the endogenous circadian oscillation adopts the exact 24-hour periodicity of the driver and also adopts a particular phase relationship to it. Entrainment therefore involves both period-control and phase-control, both of which appear to provide some selective advantage to the species.

By achieving steady-state entrainment to the Zeitgeber, an animal or plant can partition its activities into some kind of temporal order, and thereby perform behavioural and physiological activities at the "right time of the day". Many animals, for example, restrict their locomotory and feeding activity to the hours of darkness (are *nocturnal*), whereas others restrict their activity to the daylight hours (are *diurnal*); still other animals are dusk-active, or *crepuscular*. The selective advantage of this behaviour may lie in the reduction of direct competition between, say, diurnal and nocturnal species using the same food source. Examples of diurnal and nocturnal periodicity in natural and artificial light cycles are shown in figure 1.4.

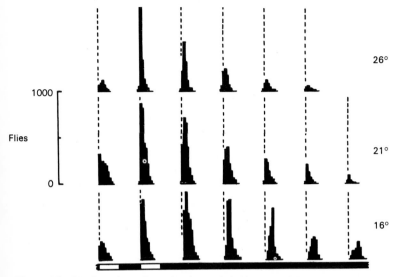

Figure 1.5   Rhythm of pupal eclosion in the fruit fly *Drosophila pseudoobscura* in a light cycle (*LD 12*:12) and then darkness (DD) at three temperatures. Note the persistence of the rhythm in the absence of the light cycle, and the small dependence of period on temperature. After Pittendrigh.

Similar examples can be quoted from the plant world. Flowers are known to open and close at certain times of the day with an accuracy that enabled Linnaeus to construct a floral clock outside his window which told him, at a glance, the time of the day by which bloom was out at a particular moment. Honey bees use a kind of circadian clock (the "time memory" or *Zeitgedächtnis*) to synchronize their visits to these opening times, or to the times when the flowers may produce nectar (Chapter 4). The selective advantage here, for plant and insect, must be to increase the efficiency of pollination and foraging, respectively.

Many species of insects emerge from their pupal cases close to dawn, when the relative humidity of the air is at its height (figure 1.5). In the case of *Drosophila pseudoobscura* it is known that emerging adults find greater success in the act of eclosion at these higher humidities, and this is assumed to constitute the selective advantage of such activity. Other insects, such as the fruit fly *Dacus tryoni*, possess a circadian rhythm of mating activity which, in steady-state entrainment, ensures that sexual activity occurs during the evening period of twilight, and thereby produces synchrony in the population (figure 1.1). Other species of the same genus are known to mate at other times of the day; thus the circadian system acts as an effective mechanism in the genetic isolation between sibling species (Tychsen and Fletcher, 1971).

In some cases the selective advantage of a biological rhythm is difficult to understand. One such example is the daily up-and-down movement of the leaves of certain plants, such as the bean *Phaseolus multiflorus*. Daily leaf movements such as these are amongst the earliest observations on circadian phenomena, and as long ago as 1729 the French astronomer De Mairan found that these movements would persist when the plants were transferred to the continuous darkness of a cellar and were therefore endogenous. Since this time many distinguished biologists have investigated this interesting phenomenon, including the plant physiologist Wilhelm Pfeffer, and Charles and Francis Darwin who devoted a whole book to *The Power of Movement in Plants* (1880). Erwin Bünning, who has studied circadian leaf movements for over 40 years, recently admitted, perhaps light-heartedly, that he used to believe that leaf movements were "only an expression of the plant's kindness towards botanists, in allowing them to discover and record circadian rhythms within the plant". Recently, however, Bünning (1971) has proposed that the adaptive value afforded by leaf movements lies in the fact that the photoperiodic receptors of plants are in the leaf epidermis, and that the upward movement of the leaves after dawn allows for the precise perception of the photoperiodically decisive twilight intervals, and the downward movement at night lessens the potentially disturbing effects of moonlight.

Another related question is what adaptive value is provided by having an endogenous oscillator to control overt behavioural events, rather than relying on direct (exogenous) responses to environmental signals, such as the changes in light intensity at dawn and dusk. The answer almost certainly lies in "preparation" and "anticipation". For example, with an endogenous oscillation, all aspects of an animal or plant's physiology are being prepared for, say, a change in the level of activity after dawn by a continuous modulation of its physiology. A mouse is in the process of "waking up" sometime before the lights go off, and does not rely on the signal itself to shake itself abruptly out of some deep slumber. An oscillatory clock also allows the honey bee to anticipate or "remember" when flowers present their nectar and, if inclement weather keeps them inactive in the hive for a day or so, they are still able to forage successfully when they finally emerge. The method of entrainment of the biological oscillator to the Zeitgeber also allows for the adoption of a particular phase-relationship which could be more difficult to achieve with an exogenous response. This is not to say that exogenous responses never occur; they do, and most biological rhythms, although endogenous, are continuously modulated under field conditions by the direct effects of the environment. The relative importance of endogenous and exogenous components varies with the species, and will be examined further in Chapter 3.

Apart from the overt behavioural rhythms discussed above, there is a wealth of covert physiological rhythms. In unicellular animals and plants, the concept of a central circadian pacemaker or "master clock", which is coupled to the light cycle on the one hand, and to a number of driven rhythms on the other, is perhaps an acceptable interpretation of the available data (Sweeney, 1969). Multicellular animals (and perhaps plants), however, seem to contain a whole "population" of circadian "driving" oscillations and an even more abundant array of driven rhythms—which, in total, comprise the *circadian system*. In steady-state entrainment to a 24-hour environmental Zeitgeber, all of the constituent oscillators achieve a mutual phase-relationship, one to another. When all Zeitgebers are experimentally excluded, however, the various components may dissociate (Chapter 2), revealing this multioscillator construction.

There is some evidence that the entrained steady state in which all the subsystems achieve a mutual phase-relationship is essential for "normal" physiological performance. For example, if plants such as the tomato are maintained in the absence of temporal cues (e.g. in constant light and temperature) they become necrotic (Hillman, 1956); if they are maintained in highly abnormal light:dark cycles they may similarly show suboptimal growth rates (Went, 1959). Amongst the insects, Pittendrigh and Minis (1972) have shown that adults of *Drosophila melanogaster* survive longer in a light cycle whose period is 24 hours than in one with a 21 or a 27-hour periodicity. Larval growth rate in the flesh fly *Sarcophaga argyrostoma* is seemingly also a function of the circadian system achieving internal synchronization (Saunders, 1972). Aschoff *et al.* (1971) have further shown that adults of the blowfly *Phormia terraenovae* subjected to weekly 6-hour phase-shift in the light cycle, either advance or delay (simulating transatlantic flights over six time zones!) showed a 20 per cent reduction in their life-span. The internal temporal organization of animals and plants is therefore clearly a matter of great importance. The discovery of "clock" mutants in *Drosophila melanogaster* (Konopka and Benzer, 1971) and the recognition of single-gene effects on the circadian period of *Chlamydomonas reinhardi* (Bruce, 1972) indicates that these phenomena are "coded for" in the genome, and are therefore every bit as much a part of the organism as its morphological organization.

Finally it must be observed that whilst some aspects of an organism's activities are "clock-controlled", many others are not. For example, eclosion of the adults of *Drosophila pseudoobscura* from their puparia is strongly coupled to the circadian clock, but puparium formation is not. It must be assumed that such couplings have evolved only where they present some selective advantage to the species concerned. To paraphrase a statement by Marshall (originally referring to the role of light in bird photoperiodism) we could say that "Time is important only in species for which it is important that it should be important".

How did biological rhythms originate?

Although the modern functional significance of many circadian and other rhythms appears clear—or at least, probable selective advantages for them can be discerned—it is more difficult to see how these oscillations might have originated, particularly so since many of the selective forces involved seem to be rather weak. However, many aspects of cellular and general physiology are oscillatory in nature. These phenomena range from high-frequency oscillations such as those involved in heartbeats and certain nervous activities to the longer periodicities involved in the cell cycle, or in the life cycle itself. It is also known that many enzyme systems in cells, or even in cell-free extracts, can produce undamped oscillations of high frequency (Pye and Chance, 1966) (figure 9.4). Indeed, Oatley and Goodwin (1971) have suggested that most aspects of cell dynamics are oscillatory, and that this is to be expected wherever temporal constraints are required to separate incompatible events in a process. At a very general level, any form of "motion" cannot proceed in one direction for ever (with the exception of evolutionary processes?) and must, at some time, return to its "starting point". It may be that the biological world, like the physical, is pervaded at all levels by oscillatory or cyclical processes. Furthermore, it is probable that the special features of circadian rhythms have been "selected" during evolution from such an oscillatory background, through the agencies of the selective forces described in the preceding section. How the accurate 24-hour periodicity was achieved, or how temperature compensation evolved, is far from clear, but a theory of strong mutual coupling between rhythms of higher frequency has been postulated as the basis of the circadian oscillation (Pavlidis, 1969). Pittendrigh (1966) has suggested that the *original* functional significance of circadian rhythms was that the reading and replication of the genetic message in eukaryotic cells became routinely executed in the dark because of some adverse effects of visible or ultra-violet radiation on essential molecular constituents of the cell. Winfree (1975), on the other hand, noting the frequent lack of discernible selective value, has suggested that the role of natural selection in adapting cells to a periodic milieu may not have been to refine a timer so much as to weed out irregular fluctuations and oscillations of maladaptive period, and to synchronize the rest. This, he believes, discourages the belief in a unique evolutionary ancestor of contemporary clocks. Nevertheless, if circadian oscillations became part of the early temporal organization of these cells, it is not surprising that they should become the pacemakers involved in so many and diverse forms of "biological clock".

The endogenous: exogenous controversy

The view presented in this book (and shared by the majority of authors) is that the timekeeping ability of organisms is *endogenous,* and derived from the physics and chemistry of their cellular physiology. Professor Frank Brown and his co-workers, however, argue that circadian and other biological rhythms derive their timing from exogenous sources or "subtle geophysical variables" not excluded in laboratory experiments. This is not to say that the endogenous "school" does not recognize exogenous effects; it does (see Chapter 3). Similarly, the exogenous "school" recognizes some sort of biological clock within the organism. The argument seems to concern the role of the environment: in the first it acts as an entraining agent, or Zeitgeber; in the second as the *source* of the periodicity. A juxtaposition of these two views is to be found in Brown, Hastings and Palmer (1970).

An unequivocal experiment to distinguish between these two alternatives would necessitate sending rhythmic organisms into a solar orbit away from all terrestrial influences which might synchronize (or cause) their rhythmicity. Despite the millions of dollars and roubles spent on space exploration, however, this experiment has apparently yet to be done. A number of rhythmic systems, including hamsters (running activity), *Neurospora* (growth rhythms), *Drosophila* (pupal eclosion), *Phaseolus* (leaf movements) and *Periplaneta* (running activity), however, have been maintained at the South Pole, on a turn-table arranged to rotate counter to the Earth's rotation every 24 hours. This experiment excluded many of the factors which might be candidates for the role of "subtle geophysical drivers", but still allowed the rhythmicity to persist (Hamner *et al.*, 1962).

Although such experiments are undoubtedly required for the unequivocal demonstration of endogeneity, the author finds that many of the data presented in this book are equally persuasive. Thus the free-running period of the biological oscillator, although often precise to within a few minutes per day, seldom *exactly* matches the dominant periods of geophysical variables, and therefore drifts out of phase with environmental time. Moreover, the period, although highly temperature-compensated (Q_{10} 0.8 – 1.2) *is* affected by temperature. The period is also adjustable by chemical treatment, notably by heavy water, although once again showing a remarkable resistance to such perturbation. Finally, stable phase-shifts and overt activity at *any* time of the solar day can be obtained, and these results seem to preclude a particular phase-relationship to an unknown and uncontrolled geophysical variable.

CHAPTER TWO

CIRCADIAN RHYTHMS:
THE ENDOGENOUS OSCILLATOR

Free-running activity in the absence of a Zeitgeber

The existence of a rhythm of behavioural or physiological activity under field conditions or in an artificial light/dark (*LD*) cycle in the laboratory tells us little about the nature of such rhythmicity. In particular, it provides no information about the relative importance of endogenous and exogenous components. In some organisms, the available evidence suggests that *direct* responses to the light or temperature cycle are more important, even dominantly so; these purely exogenous rhythms are, of course, excluded by definition from circadian rhythmicity. Other species—perhaps the majority—show exogenous effects associated with, for example, the abrupt changes of light intensity at dawn and dusk, which mask or modulate the endogenous periodicities derived from their circadian organization. Only in a strictly unentrained or "free-running" situation can this endogenous component be assessed.

When organisms as diverse as unicellular algae, flowering plants, insects, birds, and mammals are isolated and placed at constant temperature and in continuous darkness (DD)—with all other possible Zeitgebers, such as those caused by periodic human activity in the laboratory, excluded—the observed periodicity may persist, thereby revealing its endogeneity (figure 2.1). In its entrained steady state, the period of the rhythm is the *same* as that of the solar day (24 hours); in the absence of any Zeitgeber it "free-runs", and the natural period τ becomes either slightly less or slightly more than 24 hours. The value of τ is generally between 22 and 28 hours, depending on the organism. Thus the rhythm may rapidly become out of step with the solar day; for an organism whose τ is 23 hours, for example, the overt phase of the rhythm, whether it is the commencement of running activity, the discharge of spores, or the emergence of the adult insect from its puparium, would occur one hour earlier each day, and would be 180° out of phase with environmental time within two weeks.

The extreme values for τ seem to occur in plants: the rhythm of spore

19

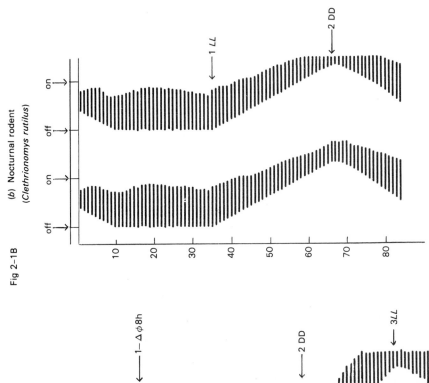

(b) Nocturnal rodent
(Clethrionomys rutilus)

Fig 2–1B

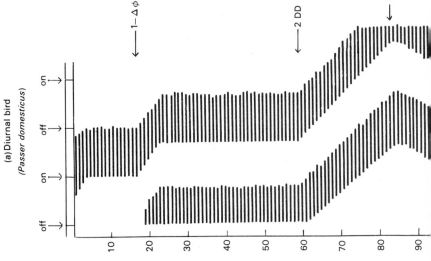

(a) Diurnal bird
(Passer domesticus)

Fig. 2.1A

Days

discharge in the alga *Oedogonium* sp., for example, shows a period of 22 hours, whereas that for the daily leaf movements in *Phaseolus coccineus* may be 27 to 28 hours. In animals the range is generally between 23 and 26 hours. The free-running period, however, is very rarely *exactly* 24 hours, and if it were one would suspect that a natural Zeitgeber associated with the solar day (whose period is 24 hours) had not been excluded.

Just as τ differs between species it also differs between individuals of the same species. In mice, for example, individual periods for locomotory activity were 25.0, 25.1, 25.3, 25.4 and 25.5 hours. In lizards τ varied from 21.1 to 24.7 hours, and human beings kept in underground "bunkers" with careful exclusion of all time cues showed natural periods ranging from 24.7 to 26.0 hours (Aschoff, 1969). Working with the flying squirrel *Glaucomys volans* kept in DD for 10 to 123 days, De Coursey (1960) showed that τ for locomotor activity varied from 22 hours 58 minutes to 24 hours 21 minutes in different individuals, but that the majority of τ-values were between 23.5 and 24.0 hours (figure 2.2).

The *accuracy* or constancy of the endogenous oscillator measured with respect to a particular overt phase, such as the onset of activity, may be quite remarkable, particularly amongst "higher" organisms such as birds and mammals. In *G. volans,* for example, the onsets of activity in DD may show a variation of only a few minutes per day; in other organisms, deviations from the average period are usually less than one hour, and frequently of the order of 15 to 20 minutes per day.

In conditions of continuous darkness and constant temperature, rhythms of activity may persist for months, as in the case of locomotor or

---

Figure 2.1   Locomotor activity rhythms of a bird (diurnal animal) and a rodent (nocturnal animal) in artificial light/dark cycles and in constant conditions (semi-schematic).

(*a*) *Diurnal animal.* After an initial series of non-steady-state cycles (delaying transients) the activity rhythm becomes entrained to the cycle of *LD 12*:12 with the onset of activity at light-on. At (1) the light cycle is phase-delayed by eight hours and the rhythm re-entrains, again after a series of transient cycles. At (2) the animal is released into continuous darkness (DD) and the rhythm "free-runs" revealing an endogenous period (τ) greater than 24 hours. At (3) the lights are turned on (LL) and τ shortens to less than 24 hours. From Menaker, M. (1971), *Biochronometry,* 315–332, Fig. 1, National Academy of Sciences, Washington.

(*b*) *Nocturnal animal.* After a series of advancing transients the activity rhythm becomes entrained to *LD 12*:12 with the onsets of activity at light-off. At (1) the animal is "released" into continuous light (LL) and the rhythm free-runs with a period (τ) greater than 24 hours. At (2) the lights are turned off (DD) and τ shortens to less than 24 hours. From Pittendrigh, C.S. (1960), C.S.H.S.Q.B., **25**, 159–184, Fig. 10.

In these records, and similar ones presented later, the horizontal black bars are produced (in the original record) by an event recorder making vertical strokes which coalesce when activity becomes intense. The records are "double-plotted" to facilitate visual inspection of the data. From various sources.

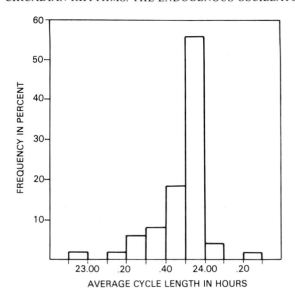

Figure 2.2   Frequency percent distribution of τ-values in DD for the flying squirrel *Glaucomys volans*. After De Coursey, P. (1960), C.S.H.S.Q.B., **25**, 49, Fig. 3.

sleep-wake activity of birds and mammals (figure 2.1), or "damp out" after a period ranging from a few weeks to a few days. This fade-out might be due to a real loss in rhythmicity, but is more probably due to an internal desynchronization of constituent subsystems, a hypothesis to which we will return later. A free-running periodicity may also persist for periods which encompass great changes in morphological organization, such as the metamorphosis from larva to pharate* adult in the fruit fly. Periodicity may also continue through periods of hibernation in certain mammals. The extreme persistence of such rhythms has led to the concept of a 'self-sustained' oscillator (Pittendrigh, 1960).

The rhythm of pupal eclosion in the fruit fly *Drosophila pseudoobscura* is one of the best known and most intensively studied examples of circadian rhythmicity, and much of what we know about any circadian system is derived from the work on this insect by Pittendrigh and his associates. This rhythm differs from those discussed earlier in this section, in that the observed event (emergence of the adult fly from its puparium) occurs only once in the life of a particular insect. However, in a mixed-age population,

---

pharate: The fully-formed adult fly lying within its puparium.

adults emerge in a series of daily pulses, each one close to dawn or to the artificial light-on stimulus provided in the laboratory (figure 1.5). With very short light components (or *photoperiods*) the eclosion peaks occur before dawn, whereas with photoperiods exceeding about six to seven hours, eclosion occurs after dawn. In a cycle of 12 hours light and 12 hours of darkness (*LD 12:*12) the median of the eclosion peak occurs about two to three hours after the onset of light.

The rhythm of pupal eclosion in *D. pseudoobscura* persists in continuous darkness (DD) with a natural period (τ) very close to 24 hours. The rhythm can also be initiated by the exposure of a culture raised since the egg stage in DD, and therefore arrhythmic, to a *single* light signal, which may be as short as 0.002 second, or by a single step-wise transfer from continuous light (*LL*) to continuous darkness (DD). The initiation of rhythmicity by these methods may be accomplished at any stage of larval or intra-puparial development, and may represent either the *initiation* of rhythmicity in each individual, or the *mutual synchronization* of individual oscillators, already running, within the population. In either case, the single pulse or the step-down in light intensity carries no information on period, and thereby constitutes positive evidence in favour of an endogenous clock.

Temperature compensation of the natural period

Although several early investigators noticed the remarkably low temperature coefficients (Q_{10}'s) of the free-running period in several rhythmic systems, it was Pittendrigh's elegant paper on the eclosion rhythm of *Drosophila pseudoobscura* (Pittendrigh, 1954) which focused attention on this phenomenon as an essential functional prerequisite for a biological clock. Pittendrigh has pointed out that not only would an oscillator which "ran faster" as the temperature rose be almost useless as a "clock", but that in the absence of such compensation, the period of the biological oscillator would at most temperatures fall outside the range of entrainment by the environmental Zeitgeber, and thus another important aspect—control of phase—would cease to function.

Many subsequent papers concerned with a wide variety of rhythmicities in an equally wide variety of organisms have demonstrated the general applicability of temperature compensation to circadian rhythms. Selected examples, including unicellular organisms, fungi, flowering plants, insects and vertebrates are shown in Table 2.1. The important observations are that whilst most physiological processes show a temperature coefficient or

Table 2.1. Selected examples showing the temperature compensation of the free-running circadian period (τ).

Organism	Rhythmic phenomenon	Temperature range, °C	Period range, τ, in hours.	Q_{10}
Euglena gracilis (alga)	Phototaxis in DD to test light	16·7–33	26·2–23·2	1·01–1·1
Oedogonium cardiacum (alga)	Sporulation	17·5–27·5	20–25	0·8
Gonyaulax polyedra (dinoflagellate)	Bioluminescence Cell division	18–25 18·5–25	22·9–24·7 23·9–25·4	0·9 0·85
Neurospora crassa (fungus)	Zonation of growth in dim red light	24–31	22–21·7	1·03
Phaseolus multiflorus (bean)	Leaf movement in DD	15–25	28·3–28·0	1·01
Leucophaea maderae (cockroach)	Locomotor activity	20–30	25·1–24·3	1·04
Schistocerca gregaria (locust)	Deposition of daily growth layers in cuticle, DD			1·04
Drosophila pseudoobscura (fruitfly)	Pupal eclosion in DD	16–26	24·5–24·0	1·02
Lacerta sicula (lizard)	Locomotor activity	16–35	25·2–24·2	1·02

Q_{10} between 2 and 3 (i.e. they more than double their rate with a 10° rise in temperature), the period of a circadian oscillator in its free-running state retains approximately the same value (i.e. possesses a Q_{10} close to 1) within a wide range of ecologically important temperatures. Generally speaking, the Q_{10}-value (calculated as the period at $(t-10)°$ / period at $t°$) is slightly greater than 1.0, implying that the frequency rises slightly with a rise in temperature. However, in some organisms, particularly the algae such as

Oedogonium cardiacum and *Gonyaulax polyedra*, Q_{10} is less than one. These values suggest that the old term "temperature independence" is not relevant to the mechanism concerned. A better term, and that used here, is "temperature compensation"; this term suggests that the apparent "independence" of temperature is gained by opposing processes with different temperature coefficients. Values of Q_{10} less than 1.0 in this view are thus regarded as examples of "overcompensation". We will return to the important but problematical question of temperature compensation in a later chapter. At this point it is important to note that the deviation from unity, however slight, is of significance because it demonstrates that the rhythm is not being entrained by an *uncontrolled* aspect of the 24-hour environmental cycle (in which case the Q_{10} would have been 1.0) and once again argues forcibly for the endogenous nature of the rhythm.

The fact that this important clock parameter is temperature-compensated does not mean that all aspects of circadian rhythmicity are so controlled. Reference to the eclosion rhythm in *D.pseudoobscura* will illustrate this point. Thus although temperature effects on period were very slight indeed ($Q_{10} = 1.02$), certain other aspects of the rhythm are markedly temperature-sensitive. For example, since the rate of development varies with the temperature, the population raised at 26° utilized a smaller number of peaks than that at 16°, but produced a larger proportion of the total adult population in each (figure 1.5). In other words, the period of the rhythm is temperature-compensated, but its "amplitude" is not.

The examples shown in Table 2.1 are all "cold-blooded" or poikilothermic animals (and plants) in which a mechanism for temperature compensation of τ would have a clear functional significance if the oscillators were to be used as "clocks". The question therefore arises as to the temperature coefficients of similar circadian periodicities in birds and in mammals—animals which have closely regulated "core" temperatures and in which temperature compensation, *a priori,* might not be expected. A number of species of mammals, however, have been subjected to deep hypothermia for varying periods, and the effect of this treatment on their rhythmicity examined. In mice (*Peromyscus leucopus*) and hamsters (*Mesocricetus auratus*), hypothermia down to 5° for 3 - 8 hours had little delaying effect on the activity rhythm, and Q_{10}'s of 1.1 to 1.3 were calculated (Rawson, 1960). Hibernating bats (*Myotis lucifugus* and *Eptesicus fuscus*) which are essentially poikilothermic in this inactive condition, were cooled in DD to a body temperature of 8 - 10° for several days (Menaker, 1959). Even in this state of induced cold torpor, endogenous rhythms (of body temperature persisted with a periodicity (τ) close to 24 hours. The evidence available from mammals thus suggests that the period of the endogenous oscillator in these normally homoiothermic animals is also temperature-compensated, and furthermore it suggests that this feature, important for chronometric purposes, is older in an evolutionary sense than the mammalian homeostatic temperature regulation.

Effects of continuous light

Many overt circadian rhythmicities in animals, particularly those concerned with motor activities such as locomotion, persist in continuous light (LL) as readily as in darkness (DD), but with their period (τ) altered. Studies like these have uncovered a generality called "Aschoff's rule" which is widely applicable to vertebrate species but finds some exceptions amongst the insects. Aschoff's rule states that the free-runing period (τ) in dark-active or nocturnal animals is longer in LL than in DD, whereas in light-active or diurnal animals τ in LL is shorter than in DD (figure 2.1). Furthermore, in LL conditions of different intensities, changes in the same direction are to be seen, depending on whether the animal is nocturnal or diurnal, τ lengthening in the former as intensity increases, but shortening in the latter. Examples of this "rule" are shown in Table 2.2. Two insects which appear to violate Aschoff's rule are also included. The mosquito *Aedes aegypti* shows a flight rhythm and is clearly day-active, but τ *lengthens* on transfer to LL, although the periodicities in both LL and DD are admittedly rather weak. Nocturnally-active ant lion larvae (*Myrmeleon obscurus*), on the other hand, show a decrease in τ upon transfer to LL, once again in seeming violation of the rule.

In birds and mammals spontaneous circadian frequency ($1/\tau$), the ratio of activity time (α) to rest time (ρ) in each daily cycle, and the overall *amount* of activity, all increase with light intensity in diurnal species, but decrease with intensity in nocturnal species. This extension of Aschoff's rule has been called the "Circadian rule". How far this phenomenon is generally applicable, or what it means in physiological terms, is not clear—apart from a superficial *a priori* expectation that day-active animals might be expected to be more active in LL than in DD, and therefore exhibit a higher frequency, and night-active animals to be more active in the dark.

The effect of continuous light on plant rhythms is very varied. In the bean *Phaseolus coccineus*, τ for leaf movement varies with quality of the light in LL. It is greatest (28.1 hours) in red light (610 - 690 nm) and least (24.7 hours) in the far red (690 - 850 nm) (Bünning, 1973). In the green alga *Oedogonium cardiacum* and the marine dinoflagellate *Gonyaulax pol-*

Table 2.2. Circadian period (τ) in continuous darkness (DD) and continuous light (LL), or as a function of light intensity (in LL). N – nocturnal animals, D – diurnal animals. An asterisk marks those examples which violate Aschoff's Rule.

Organism	Rhythmic activity	N/D	τ in DD	τ in LL	Intensity range, lx
Mus musculus (mouse)	locomotor activity	N	23–23·5	25·1–25·9	10–200
Peromyscus leucopus (mouse)	"	N	24–24·3	24·6–25·4	27–258
Glaucomys volans (flying squirrel)	"	N	23·8–23·9	24·3–24·4	0·8–5
Leucophaea maderae (cockroach)	"	N	23·3–24	24–24·7	1–270
Periplaneta americana (cockroach)	"	N	23·8	24·5	11
Teleogryllus commodus (cricket)	stridulation	N	23·5	25·3	20–35
Tenebrio molitor (flour beetle)	locomotor activity	N		24·3 25·1 26·1	0·01 2 100
Myrmeleon* obscurus (ant lion)	pit building	N	24–24·2	23·7–24	
Pyrrhula pyrrhula (bullfinch)	locomotor activity	D	26	22	
Lacerta sicula (lizard)	"	D	24	22·5–23	80–400
L. agilis (lizard)	"	D	24–24·5	22·2–23	80
Aedes aegypti* (mosquito)	flight	D	22·5	26	70

yedra, the endogenous rhythmicity is strongly damped out in DD but persists for longer periods in *LL*; this is apparently associated with a rapid depletion of reserve nutrients in prolonged darkness. In the fungi *Daldinia concentrica* and *Pilobolus* sp., on the other hand, persistence in DD is greater than in *LL* (Wilkins, 1960). In another fungus, *Neurospora crassa*, constant illumination at an intensity of 0.2 erg cm^{-2} (2×10^{-8} J cm^{-2}) (blue light, 450 nm) eliminates the circadian rhythm of growth rate within 3 days (Sargent and Briggs, 1967). In higher plants, *LL* inhibits rhythmic CO_2 emission by the leaves of *Bryophyllum fedtschenkoi* (Wilkins, 1960) and petal movements in *Kalanchoë blossfeldiana*.

Although locomotor activity rhythms in birds, mammals and some insects such as the cockroach persist in *LL*, many insect "population" rhythms, such as that governing the eclosion of the adults of *Drosophila pseudoobscura* from their puparia, are rapidly suppressed by often very low light intensities. In *D. pseudoobscura,* for example, Pittendrigh (1966) has shown that the oscillator is virtually "frozen" by light periods in excess of about 12 hours, provided the intensity is sufficiently high. When cultures are transferred to DD, the rhythm resumes with the first eclosion peak occurring 15 hours after the *LL*/DD transition and at 24-hour intervals thereafter (figure 3.10). In this case, therefore, protracted light arrests the circadian pacemaker, and on transfer to DD all of the oscillators in the population are reactivated at the same phase. Transfer of populations to *LL* of lower intensity allows some persistence of the rhythm, although damping out occurs within a few days. Winfree (1974) has since shown that photosuppression of the rhythm in *D. pseudoobscura* occurs when the light intensity reaches about 0.1 erg cm^{-2} sec^{-1} (blue light) above which arrhythmicity sets in.

The damping or fade-out of rhythms in *LL* or in DD does not necessarily mean that the underlying circadian rhythmicity has ceased: it might be due to internal desynchronization in the population, or between constituent cells or oscillators in the organism, or even between the "clock" and the overt "clock-controlled" processes. In the best investigated system (*D.pseudoobscura*), however, the motion of the circadian pacemaker itself appears to be halted in prolonged bright light.

Ontogeny and ageing effects

Most vertebrate species appear to be arrhythmic as embryos, although periodicity becomes evident soon after birth or hatching. Chickens, for example, show a rhythm of running activity immediately after emergence

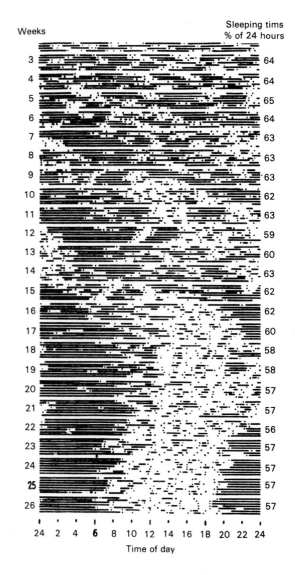

Figure 2.3 The daily incidence of sleep (black bars), wakefulness, and feeding (dots) in a human infant from birth to 26 weeks of age. Note how the initial arrhythmicity develops into a clear 24-hour pattern of activity after the 14th week. After Kleitman, N. and Engelmann, T.G. (1953), *J. appl. Physiol.*, **6**, 269-282.

from the egg. The development of such rhythmicity does not necessarily depend on exposure to an environmental Zeitgeber, birds and lizards raised from the egg in *LL* or in DD showing a spontaneous periodicity. In other species rhythmicity may develop some time after birth. In the human infant, for example, sleep-wake activity is initially at random, or nearly so, but a diurnal rhythm becomes evident after the sixth week and clearly formed after the fifteenth (figure 2.3).

In insects the question when periodicity begins has been investigated more closely. Populations of *Drosophila pseudoobscura* show a completely arrhythmic eclosion pattern when raised from the egg stage in DD, but periodicity can be initiated at any time during larval and intrapuparial development by single or repeated light or temperature signals (Pittendrigh, 1954). Zimmerman (1969) has pointed out that this arrhythmicity could be due to either (*a*) *asynchrony* of the constituent parts (i.e. organelles within cells, cells in the organism, or individuals in the population) so that all the oscillators in the population, although in motion, have scattered phases, or (*b*) a true "primary" *arrhythmicity* in which the constituent subsystems are inherited "at rest". Experiments with single temperature pulses demonstrated that the second alternative is the more likely. The stage of morphogenesis at which periodicity can be initiated from this state of primary arrhythmicity has been investigated in the pink bollworm moth *Pectinophora gossypiella* using the rhythm of egg hatch (Minis and Pittendrigh, 1968). Populations of eggs raised in DD (or in *LL*) hatch at random, but a transfer from *LL* to DD at any time after the midpoint of embryogenesis (about 132 hours from egg deposition) resulted in the generation of a distinct rhythm of egg hatch (figure 2.4). The interpretation of this result could be either (*a*) the oscillator governing the egg-hatching rhythm was not "differentiated" until this point in development, (*b*) the oscillation was present but not coupled to the light cycle, or (*c*) the overt indicator process (egg hatch) was not coupled to the clock. The first of these alternatives was favoured because the rhythm could not be initiated by a 12-hour temperature rhythm during the first half of egg development; the third alternative does not seem to be excluded, however. In the Australian fruit fly *Dacus tryoni* the rhythm of pupal eclosion can be set by a light cycle applied to the female parent (Bateman, 1955), and in the flesh fly *Sarcophaga argyrostoma* a transfer from *LL* to DD during larval development initiates rhythmic eclosion, but a similar transfer during intrapuparial development does not. In the first case, circadian rhythmicity can clearly be transferred transovarially from one generation to the next and, in the second, the coupling between the overt process (eclosion) and

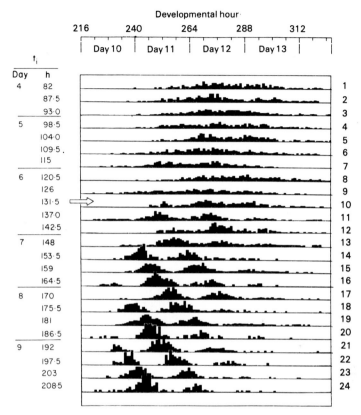

Figure 2.4 Normalized distribution of hatching in 24 populations of eggs of the pink bollworm moth *Pectinophora gossypiella*, transferred (at t_i) from continuous light (*LL*) to darkness (DD) at 5½-hour intervals from the 82nd to 208.5 hours after deposition. Note that rhythmicity develops only when the transfer occurs after the mid-point of egg development. After Minis, D.H. and Pittendrigh, C.S. (1968), *Science*, **159**, 534, Fig. 3, American Association for the Advancement of Science, Washington.

the clock, or between the clock and the photoreceptor, is broken soon after puparium formation.

In both vertebrate and insect species, the free-running period (τ), although a feature of an individual, is quite labile and may show spontaneous changes during post-embryonic development, or with age. Sometimes these changes are quite abrupt and sometimes gradual. Examples may be given from lizards, in which τ lengthened from 22.25 - 23.5 hours in very young animals to 23-25.5 hours in animals a few months old, from cockroaches and from birds and mammals. In an extensive investigation of the house sparrow *Passer domesticus,* Eskin (1971) showed that, after transfer from *LD* to DD, about half of the birds showed free-running rhythms (τ) less than 24 hours, and half more than 24 hours. Within a period of two months, however, 95 per cent of the birds had developed periods longer than 24 hours. Both short-term (4 to 6-day cycles) and long-term changes in τ were recorded, as well as the so-called "after effects" (Pittendrigh, 1960), or differences in τ attributable to various pretreatments such as different photoperiods or Zeitgeber periods to which the birds were entrained before release into DD. Pittendrigh and Daan (1974) have similarly found "ageing" effects on τ in three species of rodents (*Mesocricetus auratus, Peromyscus leucopus,* and *P. maniculatus*). In a number of animals transferred from *LD 12*:12 into DD, τ was observed to shorten as the animals aged, the natural period being significantly shorter after 14–16 months than after 5–7 months. It is interesting to note that the *direction* of these age-changes can be predicted from Aschoff's rule, which states that τ becomes longer in diurnal species on transfer from *LL* to DD, but becomes shorter in nocturnal animals. Thus, the two day-active species (*Passer domesticus* and the lizard) showed a lengthening of τ with an increasing time in DD, whereas the three species of night-active rodents showed a shortening of τ with increasing time in DD.

Dissociation of constituent subsystems

On several earlier occasions, reference has been made to the hypothesis that the circadian system in multicellular organisms consists of a number of oscillators which achieve internal synchronization when entrained by a Zeitgeber. It was further suggested that in certain conditions—continuous light and constant temperature, for example—mutual synchronization might break down and temporal organization be lost, a situation which might have deleterious consequences for some aspects of physiology (Chapter 1). Theoretical arguments in favour of the multi-oscillator construction of higher organisms may also be derived from the observation that single-celled organisms show circadian rhythmicity, so that, *a priori,* multicellular organisms literally should be "populations" of oscillators. In steady-state entrainment these constituent subsystems achieve mutual synchrony, either by independent entrainment to the Zeitgeber, by a system of mutual coupling, by coupling to "driving" oscillators, or, most probably, by a combination of all methods.

    The most direct evidence for this multi-oscillator concept of the circadian system comes from observations on the locomotor activity rhythm of

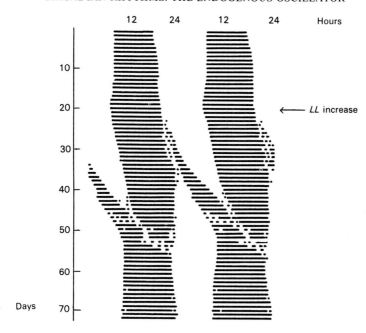

Figure 2.5   Locomotor activity rhythm of the arctic rodent *Spermophilus undulatus* in *LL*, showing the dissociation into two components, one of which adopts a lower frequency (higher τ -value) and scans the whole 24-hour record before rejoining the main band. After Pittendrigh, C.S. (1960), C.S.H.S.Q.B., **25**, 159–184.

various mammals and birds maintained in *LL* and constant temperature; in these conditions the circadian activity pattern may "split" into two components (two oscillators, or two groups of oscillators) which may free-run for a while with a different frequency. This phenomenon was first observed with the arctic rodent *Spermophilus undulatus* maintained for extended periods in *LL*. In the absence of a periodic Zeitgeber the rhythm of locomotor activity not only free-ran, but after a while two components were observed to diverge, the second running with a lower frequency (higher τ -value) but eventually rejoining the main band having scanned the full 24-hour period of the cycle (figure 2.5) (Pittendrigh, 1960). Hoffman (1971) has demonstrated a similar phenomenon in the tree shrew *Tupaia belangeri*, once again maintained in *LL*. In this case, however, splitting occurred as a function of light intensity, generally occurring when the intensity fell below 1 lx; the two components could then be made to rejoin by increasing the light intensity to values above 100 lx. Working with the starling *Sturnus vulgaris*, Gwinner (1974) has shown that splitting can be induced by the administration of testosterone. Aschoff (1969) observed essentially similar splitting phenomena in human subjects kept in an underground bunker in the absence of time cues; in this case, sleep/wake activity sometimes adopted a very low frequency ($\tau = 33.2$ hours) whilst the rhythm of deep body temperature retained its essentially circadian periodicity ($\tau = 24.9$ hours). The crux of these observations is not merely that activity splits into two components—which might be interpreted as the generation of bimodal activity from a previously unimodal pattern—but that the two components *free-run with a different frequency, or τ-value.* Therefore the two components represent two *separate* components of the circadian system which become desynchronized or dissociated in the absence of a Zeitgeber.

The multi-oscillator concept receives theoretical support from Winfree's observation that a group of weakly interacting electronic oscillators will, with a modest level of interaction, show rhythmicity due to mutual coupling, but also spontaneous dissociation similar to that observed in birds and mammals.

Similar evidence from the insects is not available, but several observations suggest that the circadian system is composed of more than one component. For example, a study of the rhythm of flight activity in the mosquito *Aedes aegypti* has indicated that both the light-on and the light-off transitions of the photoperiod have phase-setting effects and, when transferred from *LD* 4:20 into continuous darkness, both dawn and dusk peaks of activity persist at their expected times (Taylor and Jones, 1969). Separate dawn and dusk oscillators are also indicated in the photoperiodic clock of the parasitic wasp *Nasonia vitripennis* (Chapter 5).

The "gating" of population rhythms

Circadian rhythms of such events as egg hatch, pupation and eclosion from the pupa, all of which are commonly observed in insects, differ from rhythms of locomotor activity (in birds and mammals, for example) in that the former events occur only once in the life of an individual. Consequently the former require a mixed-age *population* of animals for the expression of the rhythm, whereas rhythmicity in the latter is expressed repeatedly in the individual organism. However, despite this seemingly profound difference between these types of rhythmicity, they turn out to be two aspects of the *same* phenomenon. In the individual organism, for example, an on-going oscillation dictates the time of the day at which overt locomotory activity can take place, whereas in each individual of the insect population an oscillator dictates successive "allowed zones" or *gates* through which the insect can emerge or hatch, provided that the necessary developmental steps leading up to the overt process have been completed.

The most intensively studied population rhythm is that of pupal eclosion in the fruit fly *Drosophila pseudoobscura*. In a study of the gating mechanism of this insect, Skopik and Pittendrigh (1967) released 32 separate cultures of puparia from LL into DD at different times after puparium formation; in each culture, therefore, the circadian oscillator governing the eclosion rhythm was initiated at sequentially later times relative to development. Since each culture was derived from a very narrow (5-hour) pupal collection period, the adults developing from them emerged in one, or perhaps two, peaks of eclosion (figure 2.6). Cultures 16 and 17 (released from LL into DD 96 and 102 hours after puparium formation) emerged as (female) adults in a single peak. Culture 18, on the other hand, which was released into DD 108 hours after puparium formation, showed a "split" emergence, some of the flies being able to use one gate and the others being forced to use the next. Furthermore, since the males of *D. pseudoobscura* take a longer time to complete intrapuparial development, they were forced to emerge in later gates than the females. In other words, if the developing adults were not at the "correct" morphogenetic stage for emergence at one particular gate, they were required to remain within the puparium until the next, the intervening hours constituting a "forbidden zone" for eclosion. In *D. pseudoobscura* the eclosion gates recur with circadian frequency (modulo $\tau + 15$ hours) after the LL/DD transition.

In an earlier paper (Pittendrigh, 1966) it was clearly demonstrated how a series of developmentally *synchronous* populations, in which the LL/DD transition was systematically varied, could be graphically rearranged to simulate a population of *mixed* developmental age.

The gated control of population rhythms is almost certainly ubiquitous, occurring wherever "once-in-a-lifetime" events occur in pulses every 24 hours. One more example will be given here, and that concerns the rhythm of pupation in the mosquito *Aedes vittatus* . This insect breeds in hot exposed rock pools in southern Africa, and completes its development very rapidly at the high temperatures occurring in such sites. The rock pools are also subject to periodic evaporation, and the eggs of the mosquito become dormant when dry, but may be almost immediately reactivated by a return to water, for instance when it rains. To investigate the periodicity of pupation in this insect, larval cultures were initiated by placing batches of eggs into water at three-hour intervals, eight such cultures spanning one 24-hour period. In continuous light (LL) the males formed pupae about 67 to

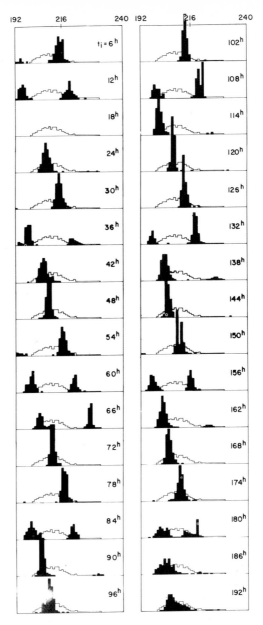

Figure 2.6 Distribution of adult emergence from 32 populations of developmentally synchronous pupae of *Drosophila pseudoobscura* transferred from *LL* to DD at 6, 12, 18, . . ., 192 hours after the beginning of pupal development. Each population is plotted against an emergence distribution in *LL* (white histogram). Note that eclosion is partitioned in some populations into two peaks or "gates", depending on the time elapsed since the *LL*/DD transition. After Skopik, S.D. and Pittendrigh, C.S. (1967), *Proc. natn. Acad. Sci.*, **58**, 1862–1869, Fig. 1.

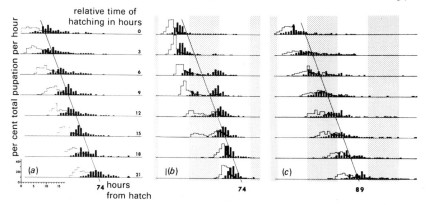

Figure 2.7 The distribution of pupation times in sequentially hatched populations of mosquito larvae. Each successively lower line represents a population hatched 3 hours later relative to the one above, and reared at 31°C under the light regime indicated by the shading. White histograms: males pupating per hour; black histograms: females pupating per hour.
(a) *Aedes vittatus* in *LL*;
(b) *A. vittatus* in *LD 12*:12;
(c) *A. aegypti* in *LD 12*:12
Note that the distribution of pupation of *A. vittatus* in *LD 12*:12 is partitioned into "gates".
After McClelland, G.A.H. and Green, C.A. (1970), *Bull. Wld. Hlth. Org.*, **42**, 951, Fig. 1, World Health Organization, Geneva.

69 hours later, and the females pupated within a further five hours. As might be expected, there was no periodicity of pupation in *LL*. In *LD 12*:12, however, evidence of gating was apparent, some of the cultures pupating in only one peak, others requiring two consecutive gates, depending on the time of hatching in relation to the *LD* cycle (figure 2.7).

Genetic control of the circadian period

Although the free-running period (τ) may be different in *LL* and in DD, may vary with the intensity of the light, and may show "spontaneous" changes during development and with age, it is clear that the *range of realizable τ-values* is characteristic of the individual organism, and is therefore probably under genetic control. Early crossing experiments with bean seedlings showing different periods in their leaf movements led Bünning to believe that this character was inherited, probably by a polygenic system. More recent experiments, however, with naturally occurring strains and induced mutants of a variety of organisms, have indicated the existence of single genes controlling the period of "clock" oscillations.

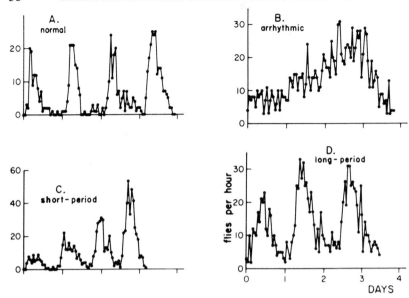

Figure 2.8   Period mutants in *Drosophila melanogaster*. The rhythms of pupal eclosion, in DD, for populations of rhythmically normal and mutant flies previously exposed to *LD 12*:12 at 20°C. After Konopka, R. and Benzer, S. (1971), *Proc. natn. Acad. Sci.*, **68**, 2112, Fig. 1.

The treatment of *Drosophila melanogaster* cultures with the chemical mutagen ethyl methane sulphonate (EMS) by Konopka and Benzer (1971) resulted in the isolation of three such period or "clock" mutants. One of them was arrhythmic; the other two were rhythmic but showed τ-values of 19 hours and 28 hours, respectively (figure 2.8). These altered rhythms were evident in both the eclosion rhythm and in a rhythm of adult locomotor activity. Since the latter was recorded using individual insects, the arrhythmicity of the first mutant was not due to a desynchronization of individuals within the population (although it could have been due to a desynchronization of separate circadian subsystems within the individuals). Furthermore, since τ for each of the period mutants was temperature-compensated, it seems most likely that the mutations were affecting the basic circadian pacemaker.

Recombination experiments with respect to morphological markers enabled Konopka and Benzer to locate the mutation on the X-chromosome. Indeed, the three mutations appeared to represent changes in the *same* functional gene or cistron—a rather surprising fact, perhaps, since one might have expected a polygenic system to control such a complex event as an eclosion rhythm. Nevertheless, results of crossing experiments indicated that the arrhythmic gene and the long- period gene were recessive to normal, whereas the short-period gene was only partially recessive. A heterozygote with one long-period and one short-period gene had a period close to normal but, when a short-period gene was opposed by an arrhythmic one, the rhythm displayed a short period. Similarly, the arrhythmic gene was "overshadowed" by a long-period one. These results suggest that the arrhythmic gene is simply inactive and the flies containing it have an "inactive" circadian oscillator.

In the unicellular alga *Chlamydomonas reinhardi,* Bruce (1972) has isolated six strains, five with a period (of a rhythm of phototaxis) close to 24 hours, and one with τ close to 21 hours. Crossing experiments have indicated that a single gene confers the long-period character.

When cultivated in a growth tube, some strains of the fungus *Neurospora crassa* (e.g. "patch") show a circadian rhythm of spore formation which is entrained by an *LD* cycle, strongly damped in *LL*, but free-runs in *DD* with a period close to 24 hours. Single genes have been isolated following treatment with chemical mutagens which affect various aspects of rhythmicity. In all cases, however, these genes appear to control the ability of the organism to *express* its rhythmicity rather than affecting the "clock" itself.

It is becoming clear that the crucial features of biological rhythms which enable them to function as a "clock" mechanism—namely, a natural period (τ) which is close to 24 hours, and which is accurate and temperature-compensated—are coded for in the genome of the organism, often by a single gene. This important observation adds considerable weight to the proposition that the biological "clock" is a product of natural selection and an intrinsic part of the organism's physiology. Other aspects of circadian rhythmicity, such as the *phase angle* adopted by the rhythm when in steady-state entrainment to the Zeitgeber, are also under genetic control; these will be examined in the next chapter, particularly with reference to the eclosion rhythm in *Drosophila pseudoobscura.*

CHAPTER THREE

CIRCADIAN RHYTHMS: ENTRAINMENT BY LIGHT AND TEMPERATURE

WHEN ISOLATED FROM ENVIRONMENTAL TIME CUES, A CIRCADIAN oscillator free-runs and reveals its natural circadian period(τ)which most usually deviates by a certain amount from 24 hours. However, in the presence of a natural solar-day Zeitgeber, such as the cycle of light intensity and temperature, the periodicity of the endogenous oscillator is adjusted (by discrete changes of different magnitude and sign) until the latter adopts the exact 24-hour period of the former. This periodicity is then maintained by similar discrete adjustments in each cycle. In this condition the endogenous oscillator is said to be *entrained* by the Zeitgeber. The most effective Zeitgeber is that of the daily light cycle, and the most important time cues are those afforded by the transitions from dark to light and from light to dark. In natural conditions, the daily photoperiod is at any one time shortening or lengthening, so that the phase angle between dawn and dusk is constantly changing. Thus, in each daily cycle, adjustments are made to both of these signals, and the overt rhythm achieves an entrained steady state in which the overt event, be it the onset of activity or the act of emergence, occurs at a particular and characteristic phase of the day-night cycle. Entrainment thus constitutes both period control and phase control.

Although cycles, pulses and stepwise changes in light intensity and temperature are the strongest Zeitgeber, other entraining agents do occur. Bird-song cycles, for example, are known to entrain avian activity rhythms, and mechanical noise in the laboratory and weak mutual interactions between individuals ("social" Zeitgebers) can also help synchronize activity.

The way in which the endogenous oscillator adjusts and becomes entrained to natural and artificial Zeitgebers constitutes the subject matter of this chapter. Most of our attention will be directed at the rhythm of pupal eclosion in the fruit fly *Drosophila pseudoobscura,* for which an empirical model for entrainment has been developed.

40

The phase response curve

One of the prerequisites for entrainment is a periodically changing sensitivity to the stimulus provided by a Zeitgeber, the signal causing adjustments (or *phase-shifts*) of different magnitude and sign depending on the *phase* at which the circadian oscillation is perturbed. The manner in which such a phase response curve is derived is best illustrated by reference to *D.pseudoobscura*.

When a mixed-age population of *D. pseudoobscura* is "released" from *LD 12*:12 or from *LL* into DD, the rhythm of pupal eclosion free-runs, producing peaks of eclosion at circadian intervals. After entry into darkness, this circadian oscillator is regarded as passing alternately through successive 12-hour half-cycles of "subjective night" and "subjective day". For the sake of convention, the various phases of the circadian oscillation are noted on the Ct (circadian time) scale which, in *D. pseudoobscura,* regards the point of entry into darkness from a light period in excess of 12 hours as *circadian time* (Ct) 12; the overt phase, or the median of pupal eclosion, then occurs about 15 hours later, or at Ct 03, and at intervals of modulo τ thereafter. This circadian time-scale should be distinguished from the Zt (Zeitgeber time) scale which describes the entraining cycle: thus Zt 00 is usually defined as the onset of the main light period, and Zt 12 the time of the light-off signal in *LD 12*:12. With steady-state entrainment to *LD 12*:12, or at the moment of transfer from a 12-hour photoperiod into darkness, the Ct and Zt-scales coincide, as defined by convention. If the photoperiod or the Zeitgeber period is changed, however, a given Ct-point may come to lie at quite a different Zt-time. The distinction between Ct and Zt is therefore important.

When the free-running oscillation in *D. pseudoobscura* is perturbed by short light signals at different circadian times, substantial phase-shifts in the steady state of the rhythm may occur, depending on the phase so perturbed (Pittendrigh, 1965) (see also figure G1 in Glossary, p. 165). For example, 15 minute light pulses applied early in the subjective night cause significant *phase delays* ($-\Delta\phi$), whereas those applied late in the subjective night and early in the subjective day cause significant *phase advances* ($+\Delta\phi$). Pulses applied between Ct 04 and Ct 12, however, have little or no effect on the subsequent steady state, and at Ct 18 (the middle of the subjective night) there is an abrupt 360° change from phase delays to phase advances. These responses to single light perturbations provide the phase response curve for the oscillation when plotted as a function of phase (figure 3.1a). For species in which τ is very close to 24 hours, the abscissa can be measured in hours on the Ct-scale, as in figure 3.1a. Where τ departs

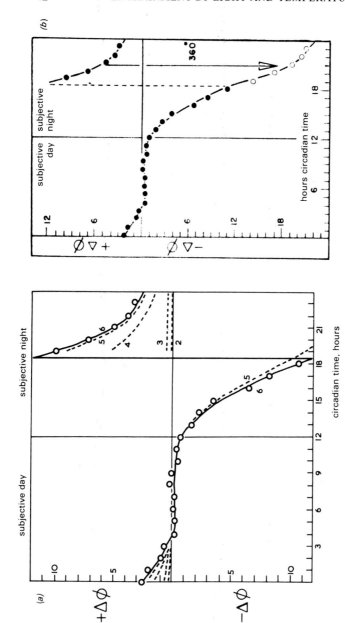

Figure 3.1(a)   The phase response curve for the *Drosophila pseudoobscura* eclosion rhythm based on 15-minute signals of white light. The solid curve and the open circles describe the steady-state phase-shifts measured on day 6 after the signal. The curves plotted as dotted lines are based on the observed shifts on days 2, 3, 4 and 5 (non-steady-state or transient cycles). Delay phase-shifts ($-\Delta\phi$) take 5 to 6 days. From Pittendrigh, C.S. (1965), *Circadian Clocks*, ed. Aschoff, J., pp. 277–297, Fig. 5, North-Holland, Amsterdam.
(b) The phase response curve in its monotonic form in which all phase shifts are shown as delays. After Pittendrigh, C.S. (1967), *Proc. natn. Acad. Sci.*, **58**, 1762–1767, Fig. 1.

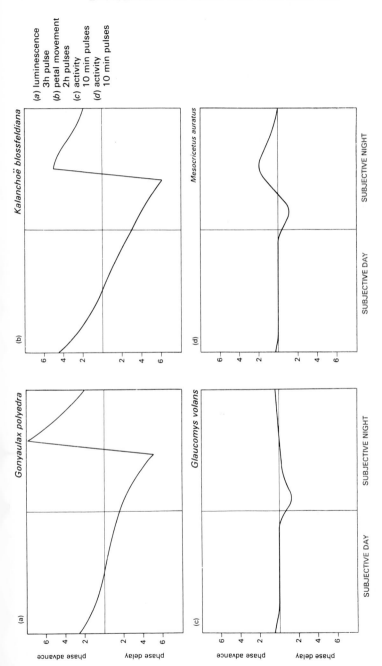

Figure 3.2 Phase response curves for four organisms.
(*a*) The marine dinoflagellate *Gonyaulax polyedra* (luminescence rhythm) subjected to 3-hour pulses of white light.
(*b*) Petal movement rhythm in *Kalanchoë blossfeldiana*, exposed to 2-hour pulses of red light.
(*c*)The activity rhythm in the flying squirrel *Glaucomys volans*, exposed to 10-minute pulses of light.
(*d*) Activity rhythm of the golden hamster *Mesocricetus auratus*, exposed to 10-minute pulses of light. (*a*) and (*b*) are strong responses giving rise to Winfree's Type 0 response curve, (*c*) and (*d*) are weak responses giving rise to Type 1 (see figure 3.11*b*). From various sources.

markedly from 24 hours, however, the Ct-scale has to be normalized to 24 hours, or the data must be plotted as a function of phase angle (0–360° of the cycle).

The modern convention for phase response curves plots delay phase shifts below the control (unperturbed) line and advances above it; it also regards the delays as having a negative sign ($-\Delta\emptyset$) and the advances as positive ($+\Delta\emptyset$) in accordance with a similar convention prevalent in the physical sciences. There are, of course, other ways of plotting response curve data: in one of these, all phase shifts are regarded as delays, to give the monotonic form (figure 3.1b); in another the interval (θ) between the resetting pulse and the mean emergence time is plotted as a function of the time (T) of the resetting pulse after an LL/DD transfer. We will discuss the second of these on p. 60; at the moment, however, we will be concerned with general properties of the phase response curve in a variety of organisms.

Response curves to single light pulses have been described for a wide variety of organisms, from unicellular algae to mammals. Some of these are shown in figure 3.2. They all show the general properties outlined for the *Drosophila pseudoobscura* case, although the "amplitude" of the phase shifts and the "areas" of the advance and delay sections of the curve may differ between species. In *D. pseudoobscura* the amplitude is great, theoretical maxima for $-\Delta\emptyset$ and $+\Delta\emptyset$ being 10 hours or more. In other rhythms, however, particularly those which govern locomotor activity in individual insects, birds and mammals, the "amplitude" for $\Delta\emptyset$ may be much less; in the hamster, for example, maximum $\Delta\emptyset$ is about 2 hours. Pittendrigh (1960) has also pointed out that the "area" under the delay section of the curve may be greater than that under the advance section of the curve for nocturnal species, but for diurnal species the opposite is true. The interpretation of these differences lies in the properties of the entrainment mechanism as described on p. 50. Lastly, it must be pointed out that the phase response curve, although characteristic for the species, is not unique; a "family" of curves for each species may be obtained, depending on the "strength" (i.e. duration and intensity) of the perturbation applied.

The circadian pacemaker and the driven rhythm

In *D. pseudoobscura*, as in all other organisms so far examined, the rhythm reaches steady state almost immediately after a single perturbation which gives rise to a delay phase-shift. In the case of an advance phase-shift, however, the system passes through several non-steady-state cycles or

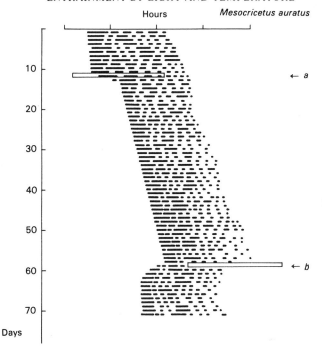

Hours   *Mesocricetus auratus*

Days

Figure 3.3 Locomotor activity rhythm of the golden hamster *Mesocricetus* free-running in DD and perturbed by 12-hour pulses of white light. At (*a*) the pulse falls early in the subjective night and causes a phase delay (−Δø) which is accomplished almost immediately; at (*b*) the pulse falls late in the subjective night and causes a phase advance (+Δø) which passes through 5–6 transient cycles before achieving steady state. After Pittendrigh, C.S. (1965), *Circadian Clocks,* ed. Aschoff, J., pp. 277–297, Fig. 3, North-Holland, Amsterdam.

transients before achieving entrainment. These effects are shown for *Drosophila* in figure 3.1a and for the hamster in figure 3.3.

Although some theoretical models for the circadian clock suggest that a single oscillator can account for these effects, transients can also be explained by the *two*-oscillator model originally proposed for the *Drosophila* case by Pittendrigh and Bruce (1959). This model has survived critical examination (in *Drosophila*) and receives experimental support from several sources. The essential feature of this model is that the physiological mechanism underlying eclosion is governed by one oscillator (the B-oscillator) which is distinct from a second and light-sensitive A-oscillator. The A-oscillator is envisaged as the central pacemaker coupled to the environmental light cycle and sensitive to the light perturbations of the kind giving rise to the phase response curve. The B-oscillator, on the other hand, is regarded as a *driven* element, coupled to the driver (A). It is furthe envisaged that the A-oscillator is "immediately" reset (i.e. perhaps withi minutes) by a light pulse, but the B-oscillator requires several cycles befor it attains a steady-state relationship to the driver (i.e. before it "catche up"); hence the series of transients. In more-modern terminology (Pitter drigh, 1967), the circadian pacemaker (A) is called the *oscillator,* and th driven element(s) (B) are regarded as *rhythms.*

The probable reality of the A and B-systems is also attested by a experiment in which "early" and "late" strains of *D. pseudoobscura* wer created from a stock culture by artificial selection for emergence tim (Pittendrigh, 1967). Fifty generations of such selection resulted in a "early" strain emerging from its puparia about four hours before the "late strain at all photoperiods between 3 and 18 hours. However, although th phase angle between the eclosion peaks and the light cycle had been altere by selection, the phase response curves for the two selected strains wer *identical* to that for the stock from which they were derived. If this result i interpreted in terms of the two-oscillator model, it may be said tha selection had altered the phase relationship between the driven rhythm (B and the driving oscillation (A), but not the phase relationship between th oscillation (A) and the light cycle.

Entrainment by one short light pulse per cycle: the primary range o entrainment.

In the next two sections we will show how the "standard" phase respons curve (15-minute pulses of white light, 100 ft-c = 1076lx) (figure 3.1a) ma be used to interpret and predict entrainment of the eclosion rhythm in *D pseudoobscura* to a variety of light cycles, according to the model propose by Pittendrigh and Minis (1964) and Pittendrigh (1965)*. It will be seer that agreement between prediction and observation is in all cases remarka bly close, a fact which supports the validity of this empirical model.

When a circadian oscillation such as that described for *D. pseudoob scura,* with a natural period τ, is entrained by a Zeitgeber with a period T the former assumes the period of the latter by undergoing discrete an apparently instantaneous phase-shifts ($\tau - T = \Delta\emptyset$) in each cycle. Wit recurrent 15-minute light pulses (i.e. one per cycle) the interval betwee them defines the "environmental period" T. For example, pulses 23 hour apart define a 23-hour environmental cycle ($T = 23$) whereas pulses 2

*This model has also proved to be applicable to the entrainment phenomenon in the house sparrow *Pass domesticus* (Eskin, 1971), a rhythm in quite a different organism and at a different level of organizatio

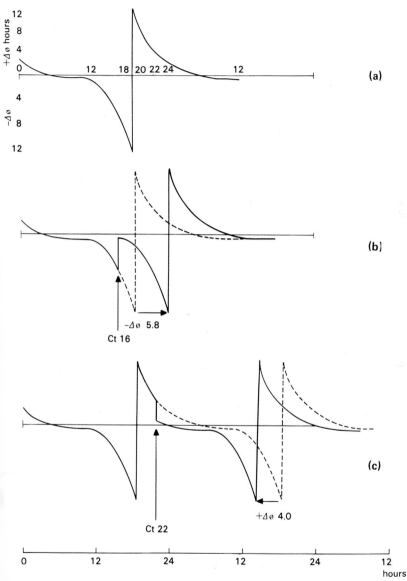

Figure 3.4 A graphical method for determining phase-shifts, based on Pittendrigh and Minis' (1964) model for *D. pseudoobscura.*
(*a*) the "standard" phase response curve (15-minute white-light pulses);
(*b*) a pulse falling at circadian time Ct 16 causes a phase delay of 5.8 hours;
(*c*) a pulse falling at Ct 22 causes a phase advance of 4.0 hours. See text for details.

(a)

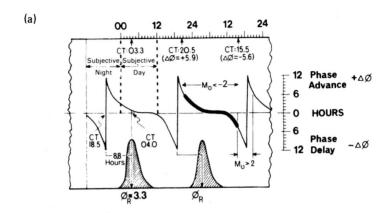

(b)

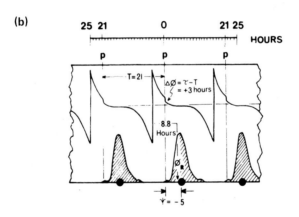

(c)

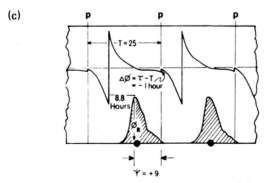

hours apart define $T = 27$. If T is thus defined by 15-minute light pulses, we can use the standard phase response curve for pulses of this duration to analyse the mechanism of entrainment, in particular at which phase point the pulse has to fall if steady-state entrainment is to occur and, indeed, if such entrainment is possible. A general observation is that, when T is shorter than τ, the oscillation phase lags the Zeitgeber, and the light pulse must fall in each cycle during the late subjective night. In this position it causes a phase advance ($+\Delta\phi$) which serves to "shorten" the period of the oscillator to that of the driver. Conversely, when τ is shorter than T, the oscillation will phase-lead the Zeitgeber, the pulse will fall in the early subjective night and generate a phase delay ($-\Delta\phi$).

Figure 3.4 shows by a graphical method how the phase response curve may be used to analyse such entrainment. The instantaneous phase-shifts are simulated by reducing the phase response curve, at the circadian time at which it is perturbed, to the control (unperturbed) value, and then moving the entire curve along this axis either to the right ($-\Delta\phi$) or to the left ($+\Delta\phi$), according to the magnitude of the phase-shift, and using the 180° phase inversion point (at Ct 18.5–19.0) as the phase reference point of the circadian oscillator. Application of this procedure to the problem of entrainment to one pulse per cycle makes it clear that, with $T = 21$, the 15-minute pulses must fall at Ct 23.3 (late subjective night) and cause an advance of $+3$ hours in each cycle if it is to correct τ (~24 hours) to T (21 hours) (figure 3.5). With $T = 25$, on the other hand, the pulses must fall at Ct 12.3 (early subjective night) to cause a phase delay of -1 hour to correct τ (~24 hours) to T (25 hours). Steady-state entrainment to T-values of the naturally occurring solar day (24 hours), or to artificial T-values greater or less than 24 hours, are thus achieved by corrections or phase-shifts of the appropriate magnitude and sign. Similar calculations using this graphical method for simulating phase-shifts will be used in later examples of the entrainment phenomenon.

Figure 3.5 Entrainment of the *D. pseudoobscura* eclosion rhythm to one 15-minute light pulse per cycle.
(*a*) the "standard" phase response curve. Entrainment is only possible when the light pulse in the steady state falls at points on the response curve where its slope (M_0) is –2 or less (heavy line).
(*b*) Entrainment by light cycles in which a 15-minute light pulse recurs every 21 hours ($T = 21$). In $T = 21$ the pulse causes a phase advance of 3 hours in each cycle, and falls at Ct 23.3 in the late subjective night.
(*c*) In $T = 25$ the pulse causes a phase delay of 1 hour and falls at Ct 12.3 in the early subjective night. After Pittendrigh, C.S. (1966), *Z. Pflanzenphysiol.,* **54,** 275–307, Fig. 8, Gustav Fischer Verlag, Stuttgart.

From this account it becomes self-evident that the "amplitude" of the phase response curve (i.e. the maximum values of $\Delta\phi$) determines the *limits* of T to which a circadian oscillator can entrain. In *D. pseudoobscura* maximum $\Delta\phi$ values approach 12 hours (figure 3.1a), but in practice entrainment is only possible when the light pulse falls on the phase response curve at points where its slope is less than –2.0. Thus the maximum utilizable $+\Delta\phi$ is 5.9 hours, the maximum utilizable $-\Delta\phi$ is 6.6 hours, and the range of entrainment is from about 18 to 30 hours. Attempts to synchronize the eclosion rhythm to Zeitgeber periods less than 19 hours, or more than 29 hours, however, result in a breakdown of entrainment. This range of realizable T-values on either side of the natural period (τ) is called the *primary range of entrainment*.

In other organisms the range of entrainment is also dictated by the extreme (advance and delay) phase-shifts which a light signal can generate, and can therefore in many cases be determined from the phase response curve. In organisms such as the flying squirrel *Glaucomys volans* and the golden hamster *Mesocricetus auratus,* the phase response curve shows smaller values of $\Delta\phi$, and the range of entrainment is consequently smaller than that for *D. pseudoobscura*. The house finch *Carpodacus mexicanus* with a low-amplutude phase response curve (± 2 hours) has a correspondingly narrow range of entrainment 21–26 hours) (Enright, 1965), but the house sparrow *Passer domesticus,* whose $\Delta\phi$ values approach 8 hours, can be entrained to Zeitgeber cycles whose periods may be as short as 17.8 hours or as long as 28.0 hours (Eskin, 1971). Amongst the insects, wide variations in the "amplitude" of the phase response curve and of the range of entrainment may also be seen. In the cockroach *Leucophaea maderae,* for example, a single 12-hour light signal commencing during the animal's subjective day caused a phase delay in the onset of locomotor activity of about 2 hours, whereas a similar pulse starting during the subjective night caused a 1-hour phase advance (Roberts, 1962). Data produced by Lohmann (1967) for this species suggested that the range of entrainment for this species was equally narrow (23–25 hours).

Entrainment by two short pulses per cycle: "skeleton" photoperiods

A "skeleton" photoperiod is formed from two (or perhaps more) pulses of light in each circadian cycle, the interval between them conveying some "information" about the duration of the simulated photoperiod. Thus, two 1-hour pulses placed four hours apart (*LD 1:4:1:18*) simulate a 6-hour photoperiod, and pulses eight hours apart (*LD 1:8:1:*14) simulate a 10-hour photoperiod. "Skeletons" may be arbitrarily divided into two types: those formed from two *equal* pulses of light (as above) are called *symmetrical skeletons*; those formed from one longer "main" photoperiod and a shorter supplementary pulse (*LD 10:2:1*:11, for example) are called *asymmetrical skeletons*. The two types present somewhat different properties and will be distinguished in the following account.

Working with the pupal eclosion rhythm in *D. pseudoobscura*, Pittendrigh and Minis (1964) reported that symmetrical skeleton photoperiods formed from two 15-minute pulses *n* hours apart were capable of simulating most of the effects of a continuous light pulse (or complete photoperiod) *n* hours in duration, provided that the separation between the two pulses was less than about 12 (or $\frac{1}{2}\tau$) hours. Not only does this observation suggest that the two pulses simulate the effects of "dawn" and "dusk" and that *both* transitions are important, but it also suggests that entrainment of circadian oscillations by such cycles is open to the kind of analysis pursued earlier with respect to one pulse per cycle.

Symmetrical skeletons (PP_s) differ from complete photoperiods (PPc) in one important respect: each complete photoperiod is a unique combination of a light period and a dark period (PPc 10 or *LD 10*:14, for example, is 10 hours of light and 14 hours of darkness); the skeleton, however, is open to *two distinct interpretations* depending on which of the two pulses we regard as the initiator and which the terminator of the simulated photoperiod. Thus, two pulses 10 hours apart (PP_s10) can also be regarded as the skeleton of a 14-hour photoperiod (PP_s14), the only difference being one of phase.

In *D. pseudoobscura*, symmetrical skeletons simulate the action of complete photoperiods almost perfectly up to about 11 hours (figure 3.6). With a 12-hour skeleton (PP_s12) the simulation is "less good", but with skeletons defining longer photoperiods (e.g. PP_s14 or PP_s18) entrainment to the longer interval is no longer possible, and the entire circadian oscillation executes a *phase-jump* to accept the shorter of the two interpretations. In *D. pseudoobscura* the general "rule" is that when any skeleton exceeds 13.7 hours, the rhythm *always* assumes a phase relation to the skeleton which is characteristic of the shorter interval (Pittendrigh, 1966).

Entrainment to such skeletons may be interpreted by reference to the phase response curve, using the same graphic method adopted for single recurrent pulses. It will be observed that the two pulses, wherever they initially strike the curve, cause discrete and instantaneous phase-shifts which "correct" the phase relationships of the oscillation to the two-point light regime until the phase advance ($+\Delta\phi$) caused by the first pulse (P_1)

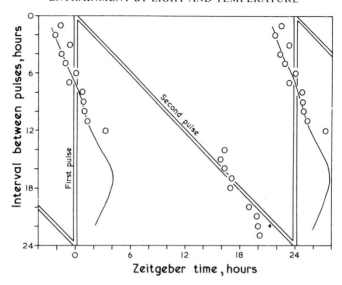

Figure 3.6   Entrainment of the *D. pseudoobscura* eclosion rhythm to two 15-minute light pulses per cycle (symmetrical "skeleton" photoperiod). When the two pulses are less than about 11 hours apart, the rhythm "acepts" the first as the "dawn" signal. When the two pulses are more than about 13 hours apart the rhythm phase-jumps to "accept" the *shorter* interval and the *second* pulse as "dawn". The open circles show the medians of the eclosion peaks; the curve shows the positions of those peaks in complete photoperiods, for comparison. After Pittendrigh, C.S. (1965), *Circadian Clocks*, pp. 277–297, Fig. 7, North-Holland, Amsterdam.

reaches an equilibrium with the phase delay $(-\Delta\phi)$ caused by the second (P_2). At equilibrium, therefore, the two $\Delta\phi$'s are of equal magnitude but of *opposite* sign, and the *net* change can be described by

$$\tau - T = (\Delta\phi_1) + (\Delta\phi_2)$$

In the case of an initial exposure to a skeleton less than about 11 hours, equilibrium is reached with the first pulse (P_1) being accepted as the initiator or "dawn" signal. For skeletons longer than 13.7 hours, however, the two pulses will cause two consecutive phase-shifts of the *same* sign until the oscillation achieves steady-state entrainment to the *shorter* interval, accepting P_2 as the initiator or "dawn" signal.

Symmetrical skeletons close to $\frac{1}{2}\tau$ have one additional property which became evident only during computer simulation in which all combinations of interval and initial phase were systematically explored (Pittendrigh, 1966). Thus, although all skeletons shorter than 10.3 and longer than 13.7 hours result in a unique phase relationship, these ambiguous regimes between 10.3 and 13.7 hours are open to *two* alternative steady states, the one being adopted depending on (1) the *phase* of the oscillation which experiences the first pulse of the entraining cycle, and (2) the value of the first *interval* experienced. These ambiguous skeletons are said to lie in the *zone of bistability*.

Asymmetrical skeletons also exert a net $\Delta\emptyset$ which is close to that obtained with a complete photoperiod of the same overall duration although, once again, a discontinuity or phase-jump occurs when the "scanning" pulse becomes too far removed from the "main" photoperiod. Entrainment to such skeletons, for the *D. pseudoobscura* case, is illustrated in figure 3.7: it is clear that when the supplementary pulse is placed early in the "night", it causes a phase delay $(-\Delta\emptyset)$ and is accepted as the terminator of the simulated skeleton photoperiod, or as a "new dusk", whereas in the latter half of the night it exerts a phase advance $(+\Delta\emptyset)$ and functions as the initiator, or "new dawn". The position of the phase jump depends on the duration of the main photoperiod, becoming later and later relative to light-on (Zt 00) as the larger component is increased. Similar phenomena are known to occur in other species, and the interpretation of the entrainment mechanism to such cycles is also relevant to the interpretation of certain photoperiodic experiments ("night interruptions") which will be examined in Chapter 5.

Frequency demultiplication and relative coordination

Circadian rhythms can entrain to environmental cycles (T) which differ from 24 hours, or from τ, provided that the extreme values are within the primary range of entrainment, as determined by the phase response curve. With T-values just ouside these extremes, entrainment no longer occurs, so that the oscillation seems to "ignore" the light regime and "free-run" across it, or aperiodicity may set in. If the periodicity of the light cycle is still further reduced, however, entrainment may reoccur, provided that T is a submultiple of τ. For example, the rodent *Peromyscus* sp. will achieve steady-state entrainment to environmental cycles such as *LD* 6:6 ($T = 12$), *LD* 4:4 ($T = 8$) and *LD* 2:4 ($T = 6$) although, of course, the entrained rhythm shows only *one* period of activity in each 24 hours (Bruce, 1960). This phenomenon is called *frequency demultiplication* and concerns the secondary ranges of entrainment.

Relative coordination was first described by Von Holst; it can be illustrated by the response of an endogenous oscillation to a potentially synchronizing Zeitgeber which has a period slightly beyond the primary

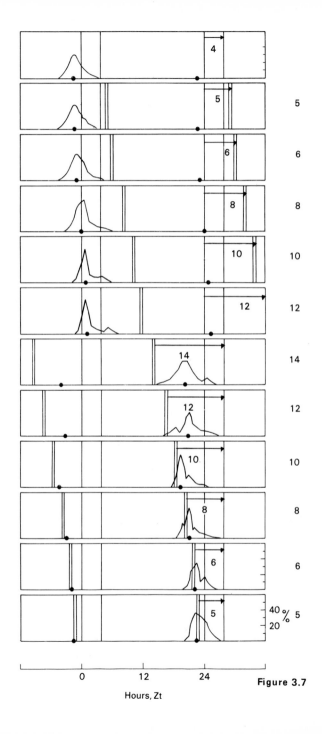

Figure 3.7

Hours, Zt

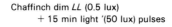

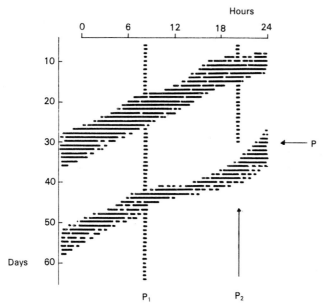

Figure 3.8 Perch-hopping activity of a chaffinch in dim *LL* interrupted by two 15-minute brighter white-light pulses (P₁ and P₂). Note (*a*) that *exogenous* activity occurs when the light pulses are on and (*b*) that, although the pulses fail to entrain the rhythm, τ becomes shorter as the activity band approaches the pulse, and becomes longer when the activity band has passed the pulse ("relative coordination"). At (p) the second pulse was discontinued, and the exogenous response abruptly ceased. After Aschoff, J. (1965), *Circadian Clocks*, pp. 95—111, Fig. 10, modified, North-Holland, Amsterdam.

range of entrainment or is, perhaps, too "weak" to enforce entrainment. Figure 3.8 shows the activity record of a chaffinch in dim *LL*, free-running with τ < 24 hours, and exposed to two 15-minute bright-light signals per cycle. Although τ does not entrain to *T*, the activity rhythm does not remain unaffected: as activity time approaches the light-pulse frequency increases (τ shortens); when the signal falls immediately after activity, frequency decreases (τ lengthens) (Aschoff, 1965).

Figure 3.7 Entrainment of the *D. pseudoobscura* eclosion rhythm to asymmetrical "skeleton" photoperiods formed from a 4-hour "main" photoperiod and a 15-minute light pulse scanning the "night". The effective "skeleton" is indicated by the arrow, and its duration by the number (of hours) beneath it. The polygons show the distributions of eclosions in each regime, and the closed circles the medians of each eclosion peak. After Pittendrigh, C.S. (1965), *Circadian Clocks*, pp. 277-297, Fig. 10, North-Holland, Amsterdam.

Spectral sensitivity and intensity thresholds for entrainment

The characterization of the relative effects of different wavelengths of light, providing an *action spectrum,* is one of the most important methods of investigating a photobiological phenomenon. In certain circumstances, for example, it may lead to the identification of the pigment molecule involved.

The action spectrum for the phase-shifting phenomenon in *Drosophila pseudoobscura* has been investigated by Frank and Zimmerman (1969), who exposed cultures to monochromatic light pulses (15 minutes) of different intensities, at either Ct 17 or at Ct 20. White-light pulses at Ct 17 are known (from the "standard" phase response curve) to generate a phase delay (–Δø) of about 8 hours, whereas a similar pulse at Ct 20 causes +Δø of about 7 hours (figure 3.1a). The magnitudes of the Δø's caused by the experimental monochromatic light pulses were therefore compared with those for the white-light control and expressed as a percentage of the latter in the oscillation's steady-state condition 7 days after the signal. Figure 3.9

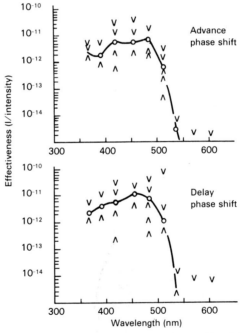

Figure 3.9 Action spectra for advance and delay phase-shifts of the *D. pseudoobscura* eclosion rhythm. 15-minute monochromatic light signals; curve connects 50 per cent phase-shifts. After Frank, K. D. and Zimmerman, W.F. (1969), *Science,* **163**, 688–689, Fig. 2. American Association for the Advancement of Science, Washington.

shows that the action spectra for $-\Delta\phi$ and $+\Delta\phi$ are essentially similar. The most effective wavelengths are those between 420 and 480 nm (blue–green); a sharp cut-off occurred with light above 500 nm (yellow–orange) and there was almost no phase-shifting caused by a broad band of intense light from the red (600 nm) to the infra-red. A very similar action spectrum was found for the initiation of the egg hatch rhythm in the pink bollworm moth *Pectinophora gossypiella* (Bruce and Minis, 1969), and for the rhythm of conidiation in the fungus *Neurospora crassa* (Sargent and Briggs, 1967). Red light is effective, however, in certain photosynthetic organisms, such as *Gonyaulax polyedra, Kalanchoë blossfeldiana* and *Phaseolus multiflorus.*

The remarkable similarity of the action spectra in *Drosophila, Pectinophora* and *Neurospora* (but not in green plants) might suggest a pigment, common to all three, and possibly of the carotenoid type as in the visual chromophore (rhodopsin). However, experiments by Zimmerman and Goldsmith (1971) with *D. melanogaster* raised on carotenoid-free diets have suggested otherwise. Fruitflies, like other animal species, are unable to synthesize their own carotenoids and derive them all from their diet, ultimately from plant sources. Therefore, if carotenoids are involved in the phase-shifting of circadian rhythms, the magnitude of $\Delta\phi$, as well as visual acuity, should be impaired in carotenoid-depleted insects. The results, however, showed that although the response of the compound eyes (containing rhodopsin) was considerably reduced, the phase-shift generated by a standard light pulse was not. Therefore, unless sufficient carotenoid is passed through the egg and used preferentially for the circadian-rhythm chromophore, these results suggest that carotenoids are *not* involved in phase-shifting of the *D. melanogaster* oscillation. The evidence that the compound eyes (or any other "organized" photoreceptor) are not involved is reviewed in Chapter 8.

Entrainment by temperature cycles, pulses and steps

In the natural environment daily cycles of light intensity are closely associated with cycles of temperature, the highest temperature often occurring in the early afternoon, and the lowest close to dawn. Not surprisingly, therefore, temperature cycles can act as important Zeitgebers, although less "strong" than changes in light. Free-running circadian rhythms in a wide variety of plants and cold-blooded animals may thus be entrained when a sinusoidal or "square-wave" temperature cycle is introduced, provided that the period of the thermal Zeitgeber is not too divergent from 24 hours (Bruce, 1960; Sweeney and Hastings, 1960). For many of these organisms a cycle with an amplitude of about 5° will suffice. In the alga *Oedogonium cardiacum,* however, a temperature cycle oscillating between extreme values only 2.5° apart is sufficient to entrain the sporulation rhythm, and in *Kalanchoë blossfeldiana* and in lizards rhythms of petal movement and locomotor activity, respectively, may be entrained by amplitudes as low as 1°. In homoiothermic animals, as might be expected, temperature cycles have less effect, the activity rhythms of various rodents, including *Glaucomys volans,* being apparently unaffected by *large* amplitude cycles (14–30°, or 15–25°). Eskin (1971), however, has recently reported synchronization to a temperature cycle of *very* large amplitude (6–38°) in the house sparrow *Passer domesticus.* The fact that such entrainment can occur in warm-blooded animals suggests that the stimulus is not one which affects the clock mechanism directly but *indirectly* via the peripheral senses.

In general it has been found that the high temperature of the daily cycle acts like the light portion of the day, and the cooler part of the cycle like the night; the oscillation normally achieves a steady-state phase relationship which reflects this. Temperature cycles are also known to "restore" rhythms which have become suppressed or "damped out" by prolonged exposure to *LL* (or DD). In *D. pseudoobscura,* for example, a temperature cycle may re-introduce rhythmicity in cultures made arrhythmic in constant light.

The entraining effects of temperature changes have been most adequately studied in *D. pseudoobscura* in which they may be interpreted in terms of phase response curves and the generalized entrainment model. Thus Zimmerman *et al.* (1968) showed that a "square-wave" cycle (12 hours at 28°, 12 hours at 20°) will entrain the eclosion rhythm in DD. They also showed that single non-recurrent temperature pulses and temperature steps (both up and down) will cause phase-shifts similar in principle to those generated by light. Temperature steps-up from 20 to 28° caused phase advances ($+\Delta\phi$) the magnitude of which depended on the phase of the oscillation so perturbed. Conversely, temperature steps-down (28–20°) caused phase delays ($-\Delta\phi$), once again magnitude depending on phase.

Just as the $\Delta\phi$'s for the "on" and the "off" signals in a light cycle (as simulated by separate 15-minute light pulses) may be summed to calculate the net $\Delta\phi$ for a light pulse, the $\Delta\phi$'s for temperature steps-up and temperature steps-down may be summed to determine the net $\Delta\phi$ for a temperature pulse, the phase at which the pulse is seen determining the net advance or delay. Furthermore, since these changes in phase were *stable*, the experiments indicated that temperature changes reset the circadian pacemaker itself (the A-oscillator) rather than some dependent system. This last proposition has been confirmed by Maier (1974) who followed a high-temperature pulse (4 minutes at 40°) with a series of 5-minute light pulses to investigate possible changes in the phase response curve. The results showed that the temperature pulse did reset the circadian pacemaker and that the phase shift so effected was "discrete" and "instantaneous".

"Conflict" between temperature and light cycles

In the natural environment, the two principal Zeitgebers (light and temperature) occur in a relationship in which the highest temperature normally occurs during the latter half of the light period, and the lowest temperature close to dawn. Both Zeitgebers serve to entrain circadian oscillations to their natural and presumably most "advantageous" phase-relationship with respect to the daily cycle. However, if the phase angle between the temperature cycle and the light cycle is altered experimentally, an interesting "conflict" between the phase-setting effects of the two Zeitgebers is observed to occur.

In *D. pseudoobscura*, for example, pupal eclosion normally occurs close to dawn. But when a sinusoidal temperature cycle (19–29°) was used in conjunction with a light cycle (*LD 12*:12), and the low point of the former moved to the "right" relative to the fixed light cycle, the peak of eclosion was observed to follow the low point until, at about 15 to 16 hours after "dawn", a discrete 180° phase-jump occurred to bring the eclosion peak forward (or back) to its original position at the onset of the light (Pittendrigh, 1960). A similar phenomenon was observed with the cockroach *Leucophaea maderae* with respect to the high point of the temperature cycle and "dusk", and with the alga *Euglena gracilis*. The results suggest that at least 180° of the 360° of conceivable phase relations to the light cycle constitute a zone of "forbidden" phase relationships.

Stopping the clock with light

There are several ways in which light can stop the motion of the circadian pacemaker. Taking the eclosion rhythm of *D. pseudoobscura* as an example, it has already been noted that protracted light of sufficient intensity "damps out" the oscillation. Pittendrigh (1966) showed that cultures transferred to darkness after any period in light longer than about 12 hours appeared to resume their "motion" at the *same* phase (Ct 12),

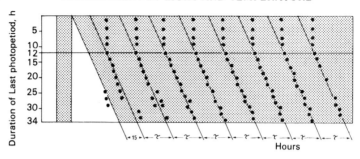

Figure 3.10 *Drosophila pseudoobscura.* The effect of varying the duration of the final photoperiod prior to releasing the cultures into a DD free-run. When the light period is greater than 12 hours the oscillation is "arrested" and begins afresh (at Ct 12) on transfer to darkness. In DD, after $L > 12$ hours, eclosion peaks (●) recur at intervals of modulo $\tau + 15$ hours. After Pittendrigh, C.S. (1966), *Z. Pflanzenphysiol.*, **54**, 275–307, Fig. 7, Gustav Fischer Verlag, Stuttgart.

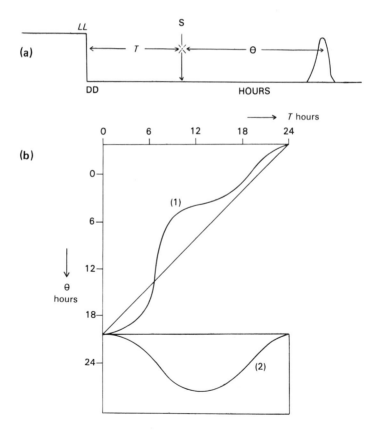

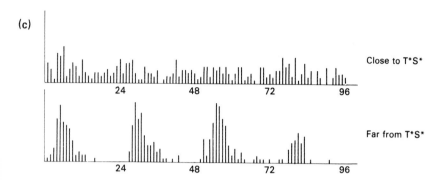

(c)

Close to T*S*

Far from T*S*

Figure 3.11 *Drosophila pseudoobscura.*
(*a*) The format of Winfree's resetting experiments. T = interval between the end of the light and the resetting pulse; S = resetting pulse; θ (cophase) = time interval + modulo τ between end of resetting pulse and the eclosion peak.
(*b*) Resetting curves at fixed stimulus duration. When $S < S^*$ (50 seconds blue light, 10 μW cm^{-2}) resetting curve (1) is of the "weak" type 1 (average slope = 1). When $S > S^*$ the resetting curve (2) is of the "strong" Type 0 (average slope = 0). The singular stimulus (S^*) occurs when Type 1 gives way to Type 0.
(*c*) Eclosion in *D. pseudoobscura* when (1) T, S is close to the singularity T^*S^* giving marked arrhythmicity, and (2) when T,S is far from T^*S^* when eclosion is strongly rhythmic. After Winfree, A.T. (1971), in *Biochronometry*, ed. Menaker, M., pp. 81–106, Figs. 2 and 6, National Academy of Sciences, Washington.

equivalent to that in a culture transferred to darkness after only 12 hours in the light (figure 3.10). The implication of this result is that photoperiods longer than 12 hours not only damp out the oscillation, but hold it (in each fly in the population) in the same fixed state. This phenomenon seems to be of wide occurrence, particularly in "population" rhythms and in the insects.

Winfree (1971) has claimed that the oscillation in *D. pseudoob-scura* can be "abolished" or placed in a completely "phase-less" state by a single short pulse of light, provided that the signal is of the correct "strength" and is applied at the right moment. Populations of flies placed in this phase-less state showed an arrhythmic pattern of eclosion, as though the clock had been stopped. The phase of the oscillator at which this critical stimulus must be applied is called the *singularity*.

Mixed-age populations were transferred from *LL* into DD to initiate rhythmicity, and then exposed to a resetting pulse (S) of blue light (10μW cm^{-2}) of different durations, and at a different time (T) after the LL/DD transition. The results were expressed in terms of the interval between the end of the light perturbation and the mean emergence time, called centroid time (θ). Plots of centroid values (θ) against T at fixed stimulus durations (S

= constant) provided phase response curves (figure 3.11) of a different pattern to those presented earlier (figure 3.1a). These resetting curves were found to fall into two distinct types (although more are theoretically possible) depending on signal "strength". Type 1, in which the average slope of the resetting response is close to 1, result from perturbations which are "weak" signals; Type 0, on the other hand, in which the average slope is 0, result from "strong" signals. In *D. pseudoobscura,* a weak signal of the defined intensity includes pulse durations of 45 seconds or less; strong signals giving rise to Type 0 occur with longer signals. (The "standard" phase response curve for this species, as shown in Fig. 3.1a, can be replotted to demonstrate that 15 minutes of white light constitute strong perturbations and give rise to a Type-0 curve).

A systematic variation of T and S showed that a critical stimulus (S^*) occurred at a point where Type-1 resetting curves give way to Type 0, and that this stimulus given a critical time (T^*) after the onset of darkness drives the oscillator into its singular state in which the clock is effectively "stopped". In *D. pseudoobscura* the singularity (T^*S^*) proves to be a 50-second pulse placed 6.8 hours after the LT/DD transition. As the values of T and S approach the singularity, there is a progressive broadening of the eclosion peaks, and at T^* S^* eclosion becomes arrhythmic (figure 3.11c). A second and third singularity occur, as might be expected, 24 and 48 hours after the first (Winfree, 1972). A slow "dark adaptation" in the system seems to occur, however, and the signal strength required at the second singularity fell from 50 seconds to about 5 seconds, and at the third singularity to about 2 seconds. There is thus more than tenfold increase in sensitivity over three days in darkness. A full account of this phenomenon and its implication in the field of clock dynamics is outside the scope of an introductory text, but may be found in Winfree's papers.

The possession of a singularity is apparently a feature of many oscillations, whether falling into the particular category called circadian or not. A similar phenomenon has been found for the circadian rhythm of petal movement in *Kalanchoë blossfeldiana,* for example, and also in the non-circadian glycolytic oscillation in yeast cells. The former rhythm, like that in *D. pseudoobscura,* concerns light as the resetting stimulus; the latter concerns the administration of oxygen to anaerobic cells which show short-period (~30 s) oscillations in NADH (nicotinamide-adenine dinucleotide).

Some direct effects of light

In the preceding sections we have seen how light (cycles or pulses) can entrain an endogenous oscillation by effecting phase-shifts which correct the endogenous period (τ) to that of the environment (T). In a large number of organisms, if not all, light has additional effects which are not to be confused with entrainment, because they modify the level or the occurrence of activity *directly*. For example, activity or some other behavioural or physiological event may be suppressed by light or by darkness. In its extreme form, the apparent activity rhythm observed under field conditions may prove to be *imposed* on the organism by the environmental light cycle. In the locust *Schistocerca gregaria,* the observed "field rhythm", with locomotor activity occurring during the night but not the day, does not persist in the laboratory either in DD or in *LL.* The rhythm also adopts a new activity pattern "instantaneously" after a shift in the phase of the environmental cycle, i.e. with no transients. Such rhythms are thought to be purely *exogenous.*

Other direct effects of light include such phenomena as "startle reactions", "rebound effects", or "masking effects" (Aschoff, 1960). Positive masking effects are those which cause an increase in activity when the light goes on; negative masking effects are those in which light inhibits activity. The former are frequently seen in diurnal animals, and the latter in nocturnal ones. An example of a positive masking effect is afforded by figure 3.8, in which the short light pulse imposes an increased activity directly on the bird, but its effects are clearly different from the partial entraining effects involved in relative coordination. Oviposition in the nocturnal moth *Pectinophora gossypiella* is almost totally suppressed by light (Minis, 1965), and this example of a negative masking effect modifies the pattern of oviposition controlled by the endogenous clock.

Exogenous light effects may involve different photoreceptors from those concerned with entrainment. In the silkmoth *Hyalophora cecropia,* for example, the photoreceptors involved in the entrainment of the pupal eclosion rhythm are situated within the brain tissue and not in the "organized" photoreceptors such as the compound eyes. Certain direct or exogenous effects associated with the effects of lights on, however, *are* mediated by the eyes (Truman, 1971). These positive masking effects were eliminated by transection of the optic nerves, or by the transfer of the brain to the abdomen, where it continued to be responsive to the entraining effects of photoperiod (Chapter 8). Transplantation of the brain with its attached eye-discs, however, led to the development of well-formed (but inverted)compound eyes, and the *immediate* or exogenous response to light was restored.

There is no doubt that such direct effects are widespread in organisms, and in many cases serve to modify endogenous rhythmicity. These effects are quite different, however, from those in entrainment. In endogenous rhythmicity the periodicity is derived from the organism's chemistry or physics, and the daily light and temperature signals serve to entrain this periodicity to that of the environment. In exogenous rhythmicity the effects of light and temperature are more direct, and the timing itself is derived from the environmental periodicity.

CHAPTER FOUR

CIRCADIAN RHYTHMS:
"CONTINUOUSLY-CONSULTED" CLOCKS

IN THE PRECEDING CHAPTERS WE HAVE EXAMINED SOME OF THE
properties of circadian rhythms controlling overt behavioural and physiological events in such a way that in steady-state entrainment to environmental Zeitgebers the overt event (such as locomotor activity or eclosion
from the pupa) occurs in a (presumably) advantageous phase-relationship
to the daily cycle of light and temperature. Such rhythms may be likened to
a "clock" in the original sense of the word*, or to a timepiece which signals
the occurrence of an event at intervals, rather like the sounding of a bell or
an alarm. In this chapter we will examine some other types of biological
clocks, similarly based on the circadian system, but in which the organism
seems to be "aware" of the time at most periods of the daily cycle. The time-
sense or *Zeitgedächtnis* of honey bees, for example, enables them to forage
at the "right time of the day" and at circadian intervals, but the insects can
also "learn" to visit several different nectar sources at different times of the
day. In the so-called *time-compensated Sun orientation* exhibited by a
variety of animals, including bees, birds and littoral crustacea, the organism can maintain a *fixed* compass direction using the Sun's azimuth as a
clue, but compensating for the sun's apparent movement through the sky
by an internal chronometer, again of a circadian nature. In both of these
types of clock the animal appears to be continuously "aware" of the passage
of time: these clocks, therefore, have been described as "continuously-
consulted" clocks, and may be likened, in terms of man-made time-pieces,
to a "watch" with hands to tell the time at any hour.

The "time-memory" or Zeitgedächtnis of honey bees
The remarkable "time-sense" or "time-memory" of honey bees (*Apis
mellifera*) has been known since the turn of the century, when the Swiss
naturalist August Forel frequently took his breakfast on the verandah of

*from *clocca,* a bell

his home and watched the bees coming to feed on the marmalade and other sweets spread on the table. He noticed that the bees tended to arrive at the same time every day, even if sweet material was not there to attract them. Forel postulated that the bees possessed a "memory for time" (Zeitgedächtnis). The German zoologist von Buttel-Reepen later described the precise times at which bees foraged in a field of buckwheat, and found that they did so only between the hours of 10.00 and 11.00 when nectar was secreted. At other times of the day, whatever the weather, the field was ignored. These observations established that bees have a remarkably accurate sense of time but, since the insects could have been responding directly to temporal cues, particularly the position of the Sun in the sky, they established little about the nature of the phenomenon. The adaptive advantage presented by the Zeitgedächtnis, however, was clear: Kleber (1935) later showed that foraging activity is closely correlated with the times at which particular species of flowers open or present nectar. Temporal synchrony between the flowers and the bees is thus obtained, and this maximizes both pollination and food collecting, respectively, with a minimum of "effort".

Modern research into the nature of the bees' Zeitgedächtnis has shown that the time measurement is a function of the circadian system. Consequently many of the properties we associate with circadian rhythms are also to be observed in this specialized aspect of bee behaviour. The daily foraging activity has been shown to persist in the absence of temporal cues, freerun with an endogenous period (τ) which is different from that of the solar day, be entrained by light-dark cycles, and have a primary range of entrainment. Moreover, the clock is undoubtedly composed of at least several components which may dissociate under certain conditions.

In Karl von Frisch's laboratory, Beling (1929) studied the bees' time memory by training bees to come to a sugar source at a particular time of the day, for several consecutive days. Each bee was individually marked whilst feeding on the sugar. During subsequent days (the "test period") the feeding dish was without sugar, but each visiting bee and its time of arrival was recorded. Beling showed that bees do indeed come back at the same time each day, with or without the "reward" of sugar, and that the bees may be "trained" to come at almost any time of the day provided that the training times were not too close together (figure 4.1). It was also notable that the same bees could attend each foraging time, and that they always visited the right *place* at the right *time*. Both Beling and Wahl demonstrated that this behaviour also occurred in the absence of obvious temporal cues, e.g. in constant light (LL), and at constant temperature and constant relative humidity, or even at the bottom of a Bavarian salt mine

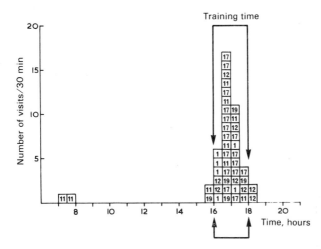

Figure 4.1 The time-memory (Zeitgedächtnis) of honey bees. The bees were "trained" to come to a sugar source at a fixed feeding position at the same time (16-18 h) during several consecutive training days. As they visited the sugar they were marked individually. On the "test day" the sugar was omitted, but the bees continued to arrive at the dish at the same time of the day. The numbers refer to the individually marked bees. After Beling, I. (1929), *Z. vergl. Physiol.*, **9**, 259-338, Springer Verlag.

where cosmic radiation was excluded. Attempts to entrain the bees' foraging activity to environmental cycles far from 24 hours failed: training times every 48 hours, for example, resulted in foraging activity every 24 hours.

The truly endogenous nature of the bees' time-sense, however, was not fully and unequivocally demonstrated until 1955 in the now classical translocation experiment performed by Renner (1955). Renner made use of two identical "bee laboratories", 7m × 3m × 3m, maintained at 28°, in *LL* (1000 1x of diffuse light), and containing a hive, feeding devices, and various figures (crosses and triangles) painted on the inside walls to aid the bees' orientation. Bees were trained to forage in one of these boxes in Paris (2° E) between 8.15 and 10.15 local time. They were then rapidly transported (by air) to an identical "bee laboratory" in New York (74° W) and tested for foraging activity on several subsequent days. Between Paris and New York the bees had been transported over 76° of longitude, or a difference of 5 hours in real local time. The rationale behind this experiment was that, if the bees relied on an endogenous circadian rhythm for time measurement, they should forage in New York exactly 24 hours after their last training time in Paris, but if they were reacting to subtle local influences they should forage in New York at the same *local* or Sun time. The result demonstrated that the former was the case: the first "morning" after arrival in New York the bees foraged at 3.00 p.m. Eastern Daylight Time, exactly 24 hours after training in Paris. Translocation in the opposite direction (west to east) produced an equivalent result. Translocation from Long Island, N.Y., to Davis, California (49° of longitude, or 3 hours in local time), this time using open fields instead of a closed environmental chamber, demonstrated that environmental Zeitgebers associated with the solar day serve to entrain the Zeitgedächtnis rhythm. On the first day following translocation, foraging activity occurred about 24 hours after the last training period on Long Island, but on the second and third days the peaks of foraging activity moved to later local clock hours by about 3 hours until the bees were foraging according to the position of the Sun at Davis.

Further evidence in favour of the endogenous circadian nature of the bees' time-memory was provided by Bennett and Renner (1963) who recorded foraging activity in constant light, constant temperature and constant humidity, for periods of 30 to 40 days. Although the bees were active at all hours of the solar day, the data suggested that the average free-running period (τ) was about 23.8 hours. Beier and Lindauer (1970) showed that τ was about 23.4 hours, that the free-running oscillation can be shifted by cycles of illumination, and that it may be entrained to Zeitgeber cycles between 20 and 26 hours. These are all properties we associate with circadian rhythms.

Like other circadian rhythms the Zeitgedächtnis clock is temperature-compensated, but may be delayed by exposure to very low temperature, chilling at 4–5°C for about 5 hours causing a 3–5 hour delay in the arrival of the bees at the feeding dish. After CO_2-narcosis (Medugorac and Lindauer, 1967) the bees turned up at the original training time *and* some time later, the delay of the second peak depending on the duration of narcosis and the concentration of the CO_2. The two peaks of foraging activity have been interpreted as evidence for more than one "component" in the clock, one of which may be stopped or delayed by the narcosis, the other continuing unaffected. The two "components" may be separate circadian subsystems, and the phenomenon may be likened to the dissociation observed with locomotor activity rhythms in mammals and birds under the influence of continuous light, or testosterone (Chapter 2).

Time-compensated orientation by the Sun, Moon and stars

Honey bees have a sense of *direction* as well as a sense of time. A bee can "learn" the direction of a food source in relation to the Sun's *azimuth* and

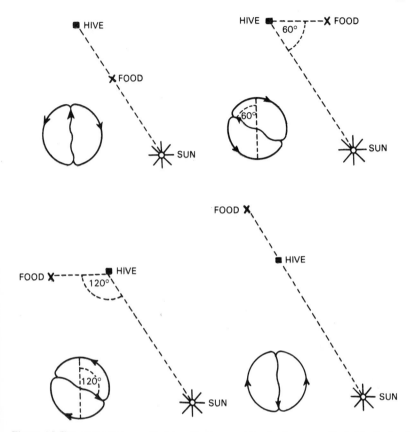

Figure 4.2 The angles between the hive, the Sun, and the food source reflected in the bees' "waggle dance" performed within the hive by returning workers. After von Frisch.

then transmit this "information" to other workers in the hive by means of the "waggle" dance, in which the angle of the forward "waggling" part of the dance to the vertical indicates the angle between the Sun's position and the food source, both in relation to the hive (figure 4.2). Other components of the dance indicate the distance of the food source. The problem faced by honey bees and all other animals which use the Sun's azimuth for maintaining a *fixed* compass direction, is to take into account the apparent movement of the Sun across the sky from East to West as the day proceeds. This compensation is accomplished by reference to an endogenous circadian clock.

Orientation by the Sun was first observed by Santschi in 1911. He showed that ants find their way back to the nest by using the Sun as a directional clue. When he shielded the ants from the *direct* rays of the Sun, but presented them with its reflected rays using a mirror, the ants predictably altered their course. Although Santschi recognized that if the ants were to use the Sun's azimuth efficiently they would need to compensate for its changing position, this possibility was considered unlikely at the time. It was almost 40 years before the work of von Frisch on bees and Kramer on starlings, published almost simultaneously in 1950, provided experimental evidence for such compensation.

Von Frisch (1950) trained bees to visit a feeding place west of the hive in the evening, then moved the hive during the night to a new site in unfamiliar surroundings. When the bees' behaviour was tested the following morning, they were found to forage to the west, even though they now had to fly away from the Sun instead of towards it as in training. A similar experiment was later performed by von Frisch and Lindauer (1954). Subsequently Meder trained bees to fly at a certain direction from the hive, then captured them at the feeding table and kept them in the dark for one or more hours. When they were released they flew unerringly in the direction they had been trained to fly in, despite the fact that the Sun had "moved" during their captivity. It has also been observed that the angle of the "waggle" dance changes during the day with the position of the Sun relative to the hive and the food source; the dancing bee thus compensates for the movement of the Sun whilst transmitting information to its hive-mates. Sometimes the bee will dance all night, and the angle of the dance changes gradually at approximately 15° per hour, indicating that the bees' inner clock can compensate for the Sun's movements, even when the Sun is on the other side of the Earth.

More progress has been made with the analysis of the Sun orientation clock in birds. It is well known that captive birds may show migratory restlessness (Zugunruhe) during the migratory season. Kramer (1950) observed that starlings (*Sturnus vulgaris*) showed such behaviour in an outdoor aviary, even if landmarks were excluded, provided that the sky was not too overcast. He used Santschi's mirror technique to demonstrate that it was the Sun which provided the directional clue, and showed that the birds were capable of maintaining a fixed compass direction, despite the movement of the Sun. Later experiments with food training, rather than Zugunruhe, and with artificial light sources, confirmed these observations.

The clock which the starling uses to compensate for the Sun's movement has proved to be circadian in nature, and is perhaps the "same" clock as that involved in the control of overt locomotor activity. Thus most of the

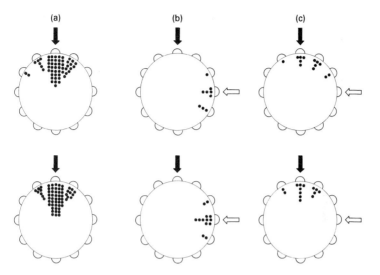

Figure 4.3 Orientation by the Sun in the starling *Sturnus vulgaris*.
(*a*)   directions chosen by two starlings in natural daylight conditions.
(*b*)   after 12-18 days in an artificial day six hours behind local time with chosen direction of orientation predictably altered.
(*c*)   8-17 days after return to natural day.
   The large circles represent the training apparatus and the 12 small semicircles on its circumference the 12 feeding dishes. Black arrows show the original training direction; open arrows the direction expected after a six-hour shift in the clock. After Hoffmann, K. (1971), *Circadian Rhythmicity,* Wageningen, pp. 175-205, Fig. 3. Centre for Agricultural Publishing and Documentation, Wageningen, Netherlands (Pudoc).

properties associated with a circadian system have been demonstrated: these include

1. phase-shifting with changes in the environmental light cycle,
2. the occurrence of a series of transient cycles after such phase-shifting,
3. free-running behaviour in the absence of a Zeitgeber.

Hoffmann found that training starlings to feed in the south in one *LD* regime, and then shifting the light cycle so that it was 6 hours later than local time, resulted in a predictable shift in the direction chosen for feeding (figure 4.3). Examination of day-by-day changes in direction after such an experiment showed that resetting is not instantaneous, but proceeds through a series of non-steady-state or "transient" cycles. Later observations (Hoffmann, 1960) demonstrated that the orientation clock may free-run under certain conditions. Birds were entrained to an *LD* cycle, then transferred to constant light for about 24 days, then back to *LD* as before.

Throughout the experiment their rhythm of locomotor activity was monitored, and at intervals they were tested for their ability to orientate themselves. After "release" into *LL*, locomotor activity "free-ran" and showed a natural circadian period (τ) of less than 24 hours, onsets getting earlier and earlier until the birds were quite out of step with their original phase. Tests for orientation showed that the direction chosen deviated

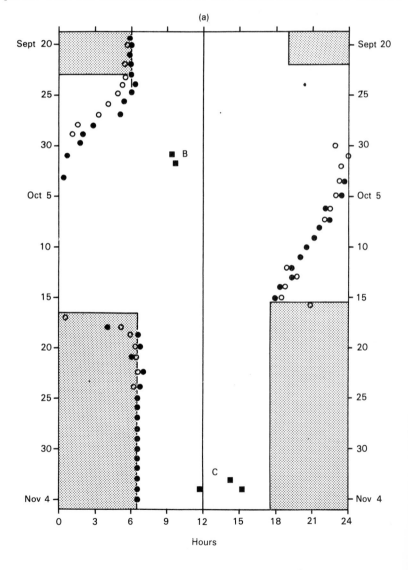

(a)

Hours

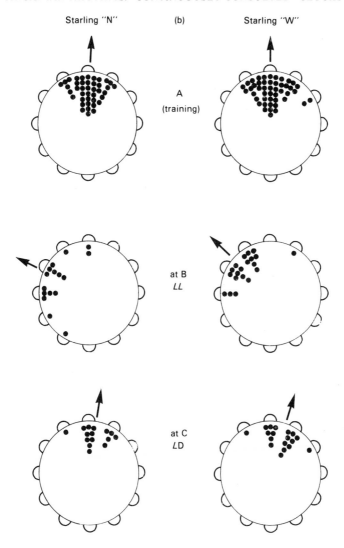

Figure 4.4   Rhythm of locomotor activity (a) and of orientation (b) of two starlings. In (a) the rhythm free-runs when 'released' into *LL* and constant temperature, onsets of activity occurring about 30 minutes earlier each day. When tested at (B) the starlings had also predictably altered their chosen feeding direction. Return to *LD* conditions brings orientation back close to its original direction. o — onsets of activity of starling N; ● — onsets of activity of starling W. Centrifugal arrows show mean direction of feeding. After Hoffmann, K. (1971), same source, as figure 4.3.

from the original training direction in a manner which could be predicted from the difference in phase between the locomotor rhythm and the natural day (figure 4.4). These observations constitute strong evidence that the orientational clock in starlings is both endogenous and circadian.

Time-compensated Sun orientation has now been described in a wide variety of animals: crustacea, spiders, fish, amphibia, reptiles and mammals, as well as in birds and insects. Notable examples include the sandhopper *Talitrus saltator,* the wolf spider *Arctosa perita,* and the frog *Acris gryllus. T. saltator* lives near the high-water mark on European beaches and uses its direction-finding abilities to maintain its position on the beach. Thus, if it is transported to higher ground, it moves down the beach towards the sea until it can bury itself in the wet sand. If it is washed out to sea, it swims up the beach to the high-water mark. The adaptive advantage of such a mechanism is obvious. As in the starling, Sun orientation in *T. saltator* contains a circadian clock element to compensate for the Sun's changing azimuth. The capacity to maintain a constant compass direction persists in constant conditions, and is independent of local time (Pardi and Grassi, 1955). The animals were also found to maintain the same angle to the Sun after transportation to Argentina, despite the fact that the Sun moves in an "anticlockwise" direction in the Southern Hemisphere.

Some organisms appear to use the Moon for maintaining compass directions, provided that the sky is clear. The sandhopper *T. saltator* is thus able to maintain its position on the beach on moonlight nights, perhaps by reference to an endogenous lunar-day clock (Papi, 1960).

The extensive literature on bird migration is mostly outside the scope of this introductory text: only those aspects relevant to time measurement will be considered. The early observations by Kramer and Hoffmann on starlings suggested that the time-compensated Sun orientational clock was involved at least in the initial stages of migration. Many birds migrate by night, however, and Sauer and Sauer (1960) found that some European warblers exhibit Zugunruhe under a starry sky, or under an artificial sky in a planetarium; orientation and migration by the stars is therefore possible. It appears that the *pattern* of stars is the clue rather than any particular star. Although that section of the night sky visible from any one place changes as the Earth revolves, it is possible that birds could find direction without "knowing" the time or the season, if the birds have some "knowledge" of star patterns. The evidence for the participation of a clock mechanism in stellar navigation is therefore equivocal.

CHAPTER FIVE

CIRCADIAN RHYTHMS: PHOTOPERIODISM

ANIMALS AND PLANTS LIVING ON THE LAND, PARTICULARLY AT HIGHER latitudes, are frequently subjected to large and sometimes violent fluctuations of temperature and humidity of a seasonal nature. These fluctuations may mean that certain times of the year are more "favourable" than others or, in an extreme case, climatic conditions at certain seasons may even threaten the organism's existence. At higher latitudes, for example, adverse winter conditions make life difficult for many organisms; similarly prolonged dry seasons in areas further south may rule out development or reproduction until the rains arrive. Only in the continuously wet tropics is continuous or homodynamic growth at all commonplace.

One of the commonest strategies evolved to counteract seasonal change is *dormancy* or, conversely, the limitation of certain events such as growth, development and reproduction to the favourable times of the year. In a very wide variety of both animals and plants, this seasonal synchrony is achieved by means of a biological clock which uses the seasonal changes in *daylength* (or night length) (Chapter 1) as an indicator of season. This type of biological time measurement—*photoperiodism*—was first described in plants (by Garner and Allard in 1920) as a mechanism controlling the initiation of flowering. Similar phenomena were then described for the control of seasonal morphs in aphids, gonadal growth in birds, and diapause induction in insects. The use of daylength as a signal to initiate seasonally-appropriate changes in metabolism is now known in many terrestrial and freshwater organisms. In plants, apart from flowering, a photoperiodic clock may control the induction and termination of dormancy in buds and bulbs, seed germination, tuber formation, vegetative growth, succulence, cambium activity, and tissue differentiation. In animals it may also control the termination of diapause in insects, migratory restlessness in birds, changes in the colour and thickness of the pelt in mammals, and the duration of delayed implantation of mammalian blastocysts. The literature on photoperiodism is equally extensive. Most of

75

this is outside the scope of this book, and we will concentrate on the evidence suggesting that photoperiodic time measurement is another function of the circadian system. Strong evidence to this effect has been obtained from flowering plants, birds and insects.

The photoperiodic clock is by no means the only mechanism which organisms possess to achieve seasonal synchrony. Many plants and animals, particularly those with a long life cycle or slow development, may also show inherent biological rhythms whose period is close to a year (circannual rhythms). These will be discussed in Chapter 6.

The photoperiodic response curve

The "standard" photoperiodic response curve is usually obtained by exposing groups of the organism being investigated to different static photoperiods (all $T = 24$ hours) and measuring the results in terms of a percentage or graded response. Figure 5.1 shows a selection of such curves in insects and birds. Frequently the response curve shows "active" development (flower formation, non-diapause development, or gonadal growth) in long days, and "inactive" processes, such as dormancy or merely vegetative growth, in short days. These organisms are called *long-day* species, and the photoperiodic response curve describes their synchrony to a favourable summer season, with dormancy supervening as the autumn approaches. Other organisms may be described as *short-day* species, with the "active" process limited to the shorter days of autumn or winter, and the "inactive" or dormant state occurring during the summer, especially when that season is unfavourably hot or dry. It must be stressed, however, that such seasonal strategies are the product of natural selection and, in the absence of selective pressures, as in the wet tropics, photoperiodic control may be absent or very weak. It may also be absent in organisms living in higher latitudes but in continuously favourable environments. Commensal insects such as the house-fly *Musca domestica,* and some pests of stored food products, for example, may be without a diapause, or a photoperiodic response. Species in which photoperiodic control is lacking may be called *day-neutral.* Still further variations of the response are possible. Some slowly-developing insects, for example, are strictly univoltine (producing one generation per year) and each generation enters diapause every winter, regardless of photoperiod. The tendency towards such univoltinism increases at higher latitudes. Other species, both plant and animal, may respond to sequences of daylength, either short to long, or long to short, or perhaps to naturally changing daylengths: these species may be able to

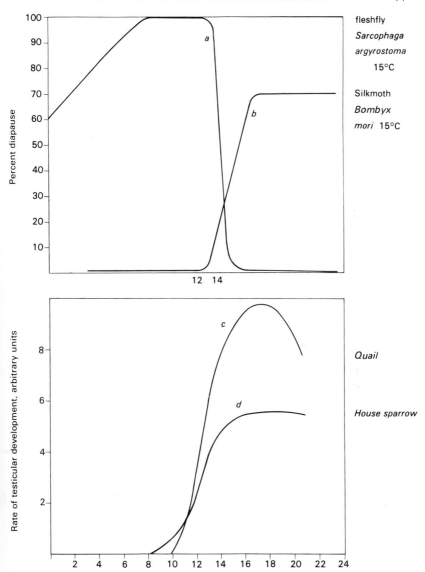

Figure 5.1 Photoperiodic response curves in insects and birds.
(*a*) induction of pupal diapause in a long-day insect *Sarcophaga argyrostoma*;
(*b*) induction of pupal diapause in a short-day insect *Bombyx mori*;
(*c*) testicular growth in the quail *Coturnix coturnix*;
(*d*) testicular growth in the house sparrow *Passer domesticus*.

distinguish the lengthening days of spring from the shortening days of autumn, and react accordingly. The photoperiodic response is therefore almost as variable as the organisms which possess it.

The most characteristic feature of the photoperiodic response curve is the so-called *critical daylength* (or critical night-length) which represents the duration of the day-time (or night-time) hours at which the organism's metabolism changes, often abruptly, from one seasonal strategy to another. This critical point varies both between species and within species, the latter according to the *latitude* from which the population was drawn. For example, the critical daylength for diapause induction in the knot grass moth *Acronycta rumicis* was found to be about 14½ hours per 24 on the Black Sea coast of the Soviet Union (43°N), whereas populations from Belgorod (50°N), Vitebsk (55°) and from Leningrad (60°) showed critical photoperiods of 16½, 18 and 19½ hours, respectively (Danilevskii, 1965). These data show that the threshold daylength changes by about 1½ hours with every 5° rise in latitude. The selective advantage of this is clear: the insects at higher latitudes compensate for the longer days of summer but earlier onset of winter with a longer critical daylength. More southerly populations, on the other hand, are able to exploit the longer favourable growing period by delaying the onset of diapause with a shorter critical daylength. The ecological "realities" of this situation were demonstrated by transferring populations from Leningrad to the Black Sea coast, and vice versa. The northern population transferred to the south entered diapause whilst the conditions for development were still favourable, whereas the southern population transferred to the north died during the first frosts of winter before preparations for winter dormancy were complete.

In both plants and animals the critical daylength is frequently temperature-compensated, or the critical daylength is less affected by temperature changes than we would expect. This, as we we have seen, is an essential functional prerequisite for accurate time measurement. However, although the time measurement itself may be temperature-compensated, other aspects of the response, including the number of flowers produced, or the proportion of insects entering diapause, may be strongly temperature-dependent. Some of these temperature effects will be examined later in this chapter (p. 95).

The use of daylength as an indicator of seasonal change has obvious advantages since, compared with seasonal changes in temperature, humidity, or the availability of plant food sources, daylength changes are almost "noise-free" and therefore predictable. Furthermore, a critical daylength well in advance of the oncoming winter allows time for the preparation of the dormant state before the first frosts arrive. In insects, this "anticipation" of diapause allows the insect to lay down reserves of fat, reduce its metabolism to a level at which it merely "ticks over", or to form special cuticular wax layers to resist desiccation.

Photoperiodic time measurement as a function of the circadian system

The possession of a critical daylength is evidence that the organism is able to "measure" either the length of the day or the length of the night (or perhaps both) in each daily cycle. The accuracy with which this time measurement is achieved is reflected by the steepness of the threshold. However, the critical daylength provides little or no information on the nature of this time measurement.Night-length or daylength measurement could, for example, be accomplished by

1. an "*hour-glass*" system set in motion *in each cycle* by the transition from light to dark, or from dark to light, or
2. by an endogenous *circadian oscillation* (or oscillations) entrained by the light/dark cycle, and measuring photoperiod by some property of its steady-state phase relationship.

The first alternative is perhaps the older, but finds support today in the writing of A. D. Lees, whose work with the aphid *Megoura viciae* will be considered later in this chapter (p. 94). The second alternative was originally suggested —for plant photoperiodism—by Erwin Bünning and has received considerable experimental support from birds and insects, as well as from plants, in the 40 years since it was first proposed. The two "alternatives", however, are perhaps not mutually exclusive. We will see, for example, that the photoperiodic clock in some organisms, although oscillatory, may measure the dark period of the cycle as if it were an "hourglass", and that both "oscillators" and "hour-glasses" may be involved in others, both plants and animals. Moreover, in those organisms in which the circadian system is clearly *somehow* involved, there is more than one way in which this may occur. This section will examine what properties the photoperiodic clock, in plants, birds and insects, has in common with overt circadian rhythmicity. Most of the evidence will be concerned with the long-day response and with the reaction to stationary photoperiods.

Bünning proposed that photoperiodic time measurement was dependent on the "endogenous diurnal rhythm" then known to provide temporal organization in plants, such as the daily up-and-down movement of the leaves. He proposed that the 24-hour period comprised two half-cycles differing in their sensitivity to light. The first 12 hours constituted a *photophil* or "light-requiring" half-cycle, and the second 12 hours a *scotophil* or "dark-requiring" half-cycle. Short-day effects were then produced when the light was restricted to the photophil, but long-day effects when the light extended, as in the summer months, into the second half (figure 5.2). This model in its most general form has since become known as "Bünning's Hypothesis". Early experiments in which the dark component of a short-day cycle was systematically interrupted by a supplementary light pulse (a night-interruption experiment) confirmed that light falling in the second half of the cycle could reverse the otherwise short-day effect. Figure 5.3 shows such results for a plant (*Kalanchoë blossfeldiana*) and a bird (*Coturnix coturnix*).

While such experiments were confined to 24-hour cycles, no evidence for circadian rhythmicity was forthcoming. However, when similar experiments were carried out with cycles with an extended "night" (e.g. with $T =$ 48 or 72 hours), periodic maxima of long or short-day effect were observed. In the long-day plant *Hyoscyamus niger,* for example, 2-hour supplementary light breaks placed systematically in the long night of an otherwise *LD*

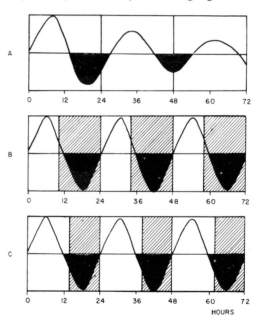

Figure 5.2 Bünning's hypothesis for the photoperiodic clock.
(*a*) the free-running oscillator in DD or *LL*;
(*b*) in short days the light does not extend into the second or scotophil half-cycle, whereas in
(*c*) (long days) it does. Originally Bunning (1960), C.S.H.S.Q.B., **25**.

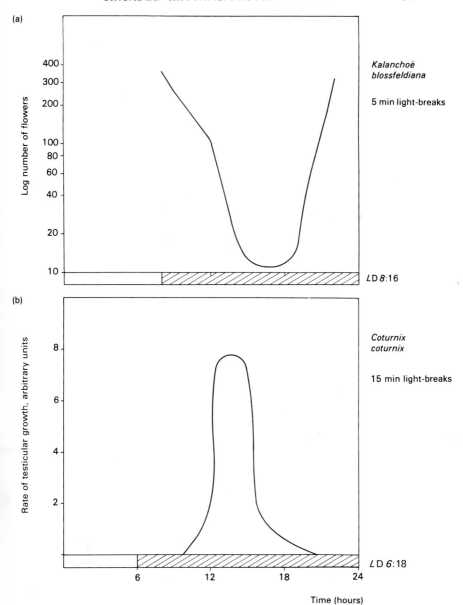

Figure 5.3   Night-interruption experiments ($T = 24$ hours) in (*a*) a short-day plant *Kalanchoë blossfeldiana* (5-minute light breaks), and (*b*) in the quail *Coturnix coturnix* (15-minute light breaks), showing a restricted period of the night which is sensitive to illumination.

9:39 cycle ($T = 48$) revealed maxima of flower induction 16 and 40 hours after the onset of the "main" light component (i.e. at Zeitgeber time, Zt 16 and Zt 40). In the soybean *Glycine max,* maintained in LD 8:64 ($T = 72$), flowering was suppressed when the light pulse fell at Zt 16, Zt 40 or at Zt 64, but was induced when the pulses fell between these times (figure 5.4). Similar results were obtained for *Kalanchoë blossfeldiana* maintained in 48-hour (bidiurnal) and 72-hour (tridiurnal) cycles. In each case the sensitivity to light pulses clearly follows a circadian rhythmicity, and in

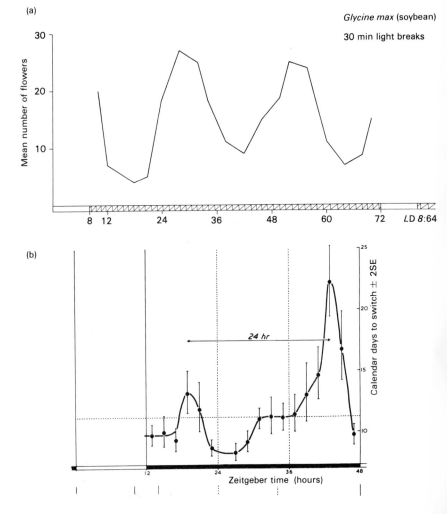

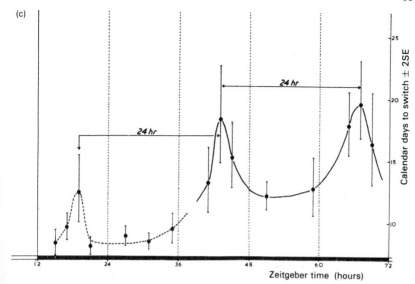

Figure 5.4 Night-interruption experiments ($T = 48$ or 72 hours) in (a) a short-day plant *Glycine max* (30-minute light breaks), (b) and (c) a long-day insect *Nasonia vitripennis* (2-hour light breaks). Note that the peaks of sensitivity to light recur at roughly 24-hour intervals in the extended "night". From Saunders, D.S. (1970), *Science*, **168**, 601–603, Figs. 1, 2, American Association for the Advancement of Science, Washington.

Kalanchoë and *Glycine* this responsiveness parallels the daily up-and-down leaf movements.

Essentially similar results have been obtained for organisms as disparate as birds and insects. In the house finch *Carpodacus mexicanus*, for example, short days (such as *LD 6*:18) are non-stimulatory, but long days (*LD 18*:6) lead to rapid testis growth and spermatogenesis. Hamner (1964) showed that birds maintained in either 24, 48 or 72-hour cycles, all with a 6-hour photoperiod, behaved as though they were at short days. Short-day responses were also obtained when an additional 1-hour light pulse was timed to fall 24 hours or 48 hours after the beginning of the "main" light component (at Zt 24 or 48), but testicular maturation did occur when the light pulses fell at Zt 12, 36 or 60. In the parasitic wasp *Nasonia vitripennis*, maintained in 48- and 72-hour cycles containing a short main photoperiod (10, 12 or 14 hours) and a 2-hour or 1-hour supplementary pulse, diapause aversion (the long-day response) was achieved when the pulses fell at Zt 19, Zt 43 and Zt 67, points which are 24 hours apart (Saunders, 1970) (figure 5.4). Pulses falling between these points induced diapause. Like the earlier

(a)

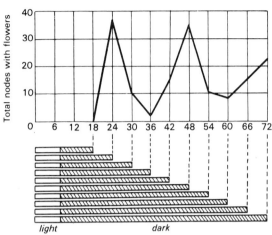

light   dark

Cycle duration hours

(b)

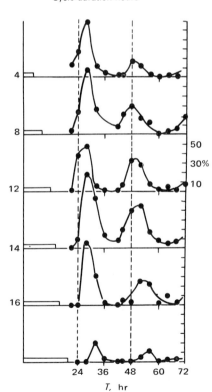

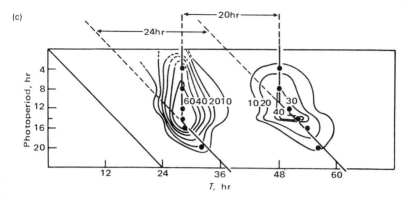

Figure 5.5 Resonance or "*T*-experiments" in (*a*) a short-day plant *Glycine max* and (*b*) a long-day insect *Sarcophaga argyrostoma*. The experimental design is shown in (*a*). In (*c*) the data for *S. argyrostoma* are replotted as a "circadian topography" to show the peaks of high diapause incidence and their relationships to the photoperiod. The contours connect points of equal diapause incidence, the solid points and the lines connecting them show the positions of the diapause maxima for each light period. The photoperiod is shown at the left as a "light wedge". (*a*) After Hamner; (*b*) After Saunders, D.S. (1973), *J. Insect Physiol.*, **19**, 1941–1954 (Figs, 5, 6) Pergamon Press, Oxford.

experiments performed with plants, these results indicate that circadian rhythmicity is somehow involved in the photoperiodic time measurement. In some other insects, however, the results were less clear-cut, or even provided evidence to the contrary. The best evidence for this contention—from the aphid *Megoura viciae*—will be reviewed later in this chapter.

A second category of experiment which has produced evidence that photoperiodism is a function of the circadian system is that devised by K. C. Hamner and his associates, and originally applied to plants. In this type of experiment (a "*T*-experiment"), the organism is exposed to a range of cycles, each containing a light component of the same length, but coupled to a dark period between, say, 12 and 60 hours, to provide abnormal cycles from less than 24 hours up to $T = 72$, or longer. In the Biloxi variety of soybean (*Glycine max*), which is a short-day plant, maxima of flower formation were observed at $T = 24$, 48 and 72, and minima at $T = 36$ and 60 (figure 5.5). Similar results have since been obtained for the house finch *Carpodacus mexicanus* (W.M. Hamner, 1963), the Japanese quail *Coturnix coturnix,* and for two species of sparrow, *Zonotrichia leucophrys* and *Z. atricapilla.* Amongst the insects, examples are afforded by the flesh-fly *Sarcophaga argyrostoma* and its parasite *Nasonia vitripennis* (Saunders, 1974). A selection of these results, all for long-day species, are presented in figure 5.5. Once again, however, attention should be drawn to some exceptions, notably the aphid *Megoura viciae,* which will be considered later.

Some more explicit models

Results from "*T*-experiments" and "night-interruption experiments" using long-period cycles may be interpreted as showing that a particular light-sensitive phase recurs with circadian frequency during the long dark period. When $T = 24$, according to Bünning's hypothesis, long-day responses occur when this phase point is illuminated, and short-day responses when it is not. C. S. Pittendrigh and his associates at Princeton and Stanford Universities, however, have pointed out that circadian rhythmicity might have more than one role in photoperiodic time-measurement, if it is involved at all, and they and other groups of workers have proposed several alternative, or more explicit models to account for the phenomenon. Pittendrigh (1972) has suggested three possible types of clock:

1. *external coincidence*
2. *internal coincidence*
3. a less explicit formulation in which circadian rhythmicity is not necessarily *directly* involved in time measurement, but the time-measuring function of the clock, whatever it is, is affected by what he calls *"circadian resonance"*.

The first two will be examined in this section. The *external-coincidence model* (Pittendrigh and Minis, 1964), which is an explicit version of Bünning's general hypothesis, was broadly based on the known properties of entrainment exemplified by the *Drosophila pseudoobscura* eclosion rhythm (Chapter 3), and on certain night-interruption experiments performed by Adkisson (1964) on the pink bollworm moth *Pectinophora gossypiella.* It proposed that light must have *two* roles:

1. *entrainment*
2. the photoperiodic *induction* caused by light falling on a light-sensitive or photo-inducible phase (ϕ_i).

The external-coincidence model, therefore, was so-named because of this coincidence, in time, between ϕ_i and an external (= environmental) light. The authors stressed that any attempt to analyse the photoperiodic clock must take into account this dual role of light, and if the two roles were experimentally separable, the clock would be open to analysis using the *Drosophila* entrainment model based on the phase response curve (Chapter 3).

In order to understand the external-coincidence model it is necessary first to examine Adkisson's data for night interruptions in *P. gossypiella.*

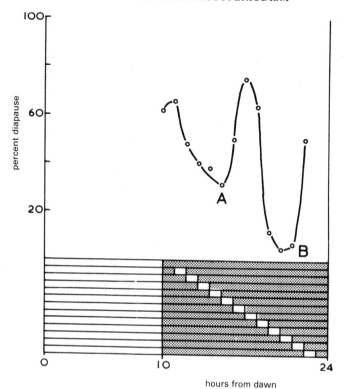

Figure 5.6 Night-interruption experiments ($T = 24$ hours) with the long-day insect *Pectinophora gossypiella,* showing the *two* points of long-day response or diapause inhibition (A and B), typical of most insect species. After Adkisson.

These experiments were similar in design to those shown in figure 5.3, but produced not one, but *two* peaks of long-day (diapause-averting) effect (figure 5.6). Such bimodal responses are now known to be widespread in the insects. Pittendrigh and Minis interpreted these results by comparing them with the responses of the *D. pseudoobscura* eclosion rhythm to similar experimental protocols, called asymmetrical skeleton photoperiods (figure 3.7). They pointed out that only one of the peaks of diapause aversion could represent the position of the photoinducible phase (ϕ_i), and that the two peaks are the result of a discontinuity or phase-jump in the entrainment phenomenon. (For a variety of reasons to be discussed later, they considered that ϕ_i must lie at the second of these two peaks). Thus pulses falling in the early part of the night act as a "new dusk" and cause phase delays until ϕ_i is brought into coincidence with the "dawn" transition of the main photoperiod, and long-day responses are produced. Pulses falling in the second half of the night, on the other hand, act as a "new dawn" and cause phase advances until the light pulse illuminates ϕ_i directly, producing the second of the long-day peaks. The point of "insensitivity" in the middle of the night represents the position of the phase-jump.

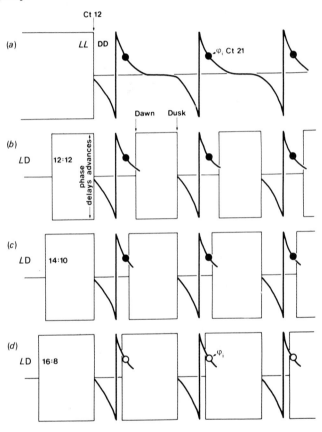

Figure 5.7 The external coincidence model for the photoperiodic clock. (*a*) the oscillator (shown here as the *Drosophila pseudoobscura* phase response curve) commencing at constant phase (Ct 12) after transfer from *LL* (or a light period longer than 12 hours) into darkness. In (*b*) (*LD 12*:12) and (*c*) (*LD 14*:10) the photoinducible phase (*ø*$_i$) falls late in the night, is not illuminated; hence short-day (or long-night) responses occur.

In (*d*) (*LD 16*:8) on the other hand, the "dawn" transition extends backwards into the preceding night to illuminate *ø*$_i$ and long-day (or short-night) responses occur. After Saunders, D.S. (1975*a*), *J. comp. Physiol.*, **97**, 97–112, Fig. 2, Springer Verlag.

In a later paper, Pittendrigh (1966) pointed out that the oscillation in *D. pseudoobscura* and in *P. gossypiella* appears to cease its motion in photoperiods longer than about 12 hours (Chapter 3), but restarts at the same (or constant) phase on transfer to darkness. Consequently, in light/dark cycles ($T = 24$) containing a photoperiod in excess of 12 hours, the oscillation is always at the same phase at the dusk transition, but the "dawn" transition effectively "moves backwards" into the preceding night, as the photoperiod increases, until it coincides with $ø_i$ (figure 5.7). In all these photoperiodic cycles the second peak of diapause aversion thus represents the phase of $ø_i$; in *P. gossypiella* it is 9 to 10 hours after dusk, an interval which constitutes the "critical night-length".

The external-coincidence model, in a slightly modified form, accounts for most aspects of the phenomenon in the flesh fly *Sarcophaga argyrostoma* (Saunders, 1975a). There are also some experiments with birds which can, perhaps, be explained only by such a model. In one of these, Menaker and Eskin (1967) entrained the locomotor activity (perch hopping) rhythm in sparrows (*Passer domesticus*) to 14 hours of dim-green light per day. Green light of this intensity, although sufficient for entrainment, was insufficient to bring about the testis growth observed in 14 hours of white light. An additional short pulse of white light (75 min) was then placed either at the beginning or at the end of the 14 hours of green, in such a way that the additional pulse had no modifying effect upon the pattern of entrainment. Only in the latter position was testicular development induced. This experiment successfully separated the two roles of light (entrainment and induction) and, as a test for external coincidence, appears to verify the model. In other species, however—including *P. gossypiella*, for which it was initially devised—the external coincidence model has proved to be inappropriate (Pittendrigh and Minis, 1971).

In *internal coincidence*, light has only one role, that of entrainment, and induction is a function of the internal phase-relationships of constituent oscillations. It was first proposed in a general form by Pittendrigh (1960) who noted that a change in photoperiod was also a change in the phase-angle between dawn and dusk. In 1966 Tyshchenko proposed that the photoperiodic clock (in insects) comprised two oscillators, each of circadian periodicity, one phase set by the dawn transition of the photoperiod, the other by dusk. As the photoperiod changed, the phase angle between the two oscillators would also change, and induction or non-induction would occur when "active" phase points of the two oscillators either coincided or failed to coincide (figure 5.8). Support of a general nature for models of this type comes from observations that the circadian system in a

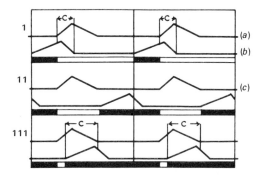

Figure 5.8   Tyshchenko's two-oscillator model for the photoperiodic clock (internal coincidence). (*a*) the oscillator phase set by "dawn"; (*b*) the oscillator phase set by "dusk"; (*c*) the times of coincidence between the two oscillators, occurring at either long daylength (I) or at very short daylength (III). In natural short daylengths (II) coincidence does not occur and diapause supervenes. Redrawn from Danilevskii *et al.* (1970), *Ann. Rev. Ent.,* **15**, 201–244, Fig. 6, Annual Reviews Inc., Palo Alto, California.

variety of organisms consists of a "population" of oscillators, some of which may be coupled to the light cycle by the dawn signal and others by the dusk.

Experimental evidence in favour of some sort of internal coincidence has come from "*T*-experiments" with the parasitic wasp *Nasonia vitripennis,* the females of which produce diapausing progeny (larvae) when raised at short daylength, but continuously-developing progeny at long daylength (Saunders, 1974). Figure 5.9 shows that the peaks of short-day effect (high diapause) produced in *T*- or resonance experiments were repeated with a circadian frequency in the long "night", but that the "ascending slopes" and the "descending slopes" of these maxima showed distinctly different relationships to the photoperiod. The "ascending slopes", for example, were clearly obtaining their principal time cue from "dusk", whereas the "descending slopes" were obtaining theirs from "dawn"; consequently the mutual phase relationship between these components altered with changes in the photoperiod, and the diapause maxima became broader or narrower. The "descending" and "ascending" slopes were identified as manifestations of the dawn and dusk oscillations in Tyshchenko's model.

If the photoperiodic clock in *N. vitripennis* is of the internal-coincidence type, one simple test is available. Since light has only a single role—entrainment—and entrainment by a light cycle in a variety of circadian rhythms can be duplicated by a temperature cycle (Chapter 3), it should be possible to simulate the effects of daylength by using daily *thermo*periods

Fig. 5.9(a)

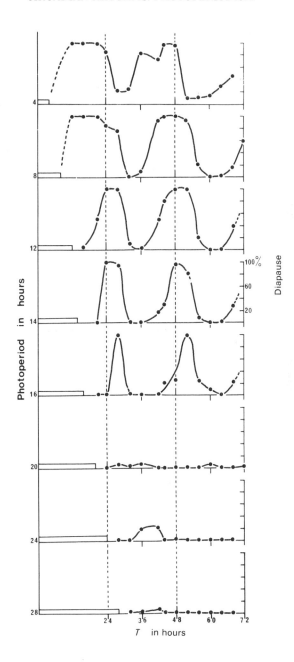

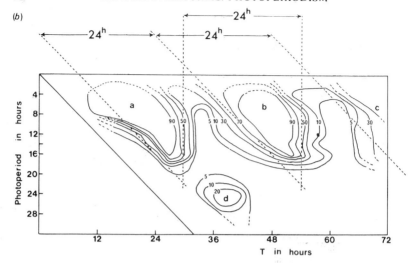

Figure 5.9   Resonance or "*T*-experiments" with the parasitic wasp *Nasonia vitripennis*. In (*b*) the data are redrawn as a "circadian topography" (see figure 5.5 for explanation). The "ascending slopes" and "descending slopes" of the diapause maxima appear to take their principal time cues from "dusk" and "dawn" respectively, and may represent manifestations of the two oscillators in the photoperiodic clock (figure 5.8). The photoperiod is shown at the left as a "light wedge". After Saunders, D.S. (1974), *J. Insect Physiol.*, **20**, 77–88, Figs. 1, 2, Pergamon Press, Oxford.

in the *complete absence of light*. This has proved possible by exposing females of the wasp, raised from the egg stage in complete darkness, to daily temperature cycles consisting of variable periods at 13° or 23°C. This experiment (figure 5.10) showed that when the daily thermoperiod (the number of hours per day at the higher temperature) was less than about 13 hours, nearly all of the wasps produced diapausing offspring, but when the thermoperiod was greater than 13 hours the offspring developed without a diapause (Saunders, 1974). Not only does this experiment show that temperature may be substituted for light in "photo-periodism", but it also rules out—for *N. vitripennis*—any model for the clock which includes an inductive, as opposed to a purely entraining, function for the environmental light cycle. It does not, of course, exclude a modified form of Bünning's hypothesis which includes a particular *temperature*-sensitive phase-point.

Hour-glasses in photoperiodic time measurement

Although the circadian system in plants, birds and some insects is clearly involved in the photoperiodic clock, the hypothesis that time measurement is accomplished *in some species* by an "hour-glass" type of mechanism should not be excluded. Results of experiments in plants and in insects have suggested that oscillators *and* hour-glasses may be involved, that oscillators may act like hour-glasses in certain circumstances, or that hour-glasses may be solely responsible. Examples from the insects will be examined here.

Pittendrigh (1966) pointed out that since the oscillation in *D. pseudoob-scura* and *P. gossypiella* is effectively "damped out" by photoperiods in excess of about 12 hours, and is held at a constant phase (equivalent to Ct 12) until darkness resumes, it is (as depicted in figure 5.7) measuring the dark period as if it were an "hour-glass". Experimental evidence for such a situation has been obtained in *Sarcophaga argyrostoma* (Saunders, 1975a). In this insect the photoperiodic "oscillator" can only reveal its circadian nature in cycles with a very long "night". In 24-hour cycles with a photoperiod greater than 12 hours (i.e. in the region where photoperiodic control is important) the oscillation commences its "motion" anew each night at the same phase (regardless of the length of the photoperiod) and measures a critical night-length (~ 9 hours) before being reset by the next light period.

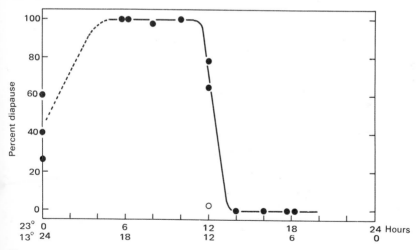

Figure 5.10   The effect of a daily temperature cycle (23°C/13°C) on the induction of larval diapause in *Nasonia vitripennis,* showing the sharp discontinuity (critical thermoperiod) between short and long thermoperiods. The experimental wasps were maintained in continuous darkness from the egg stage. The open circle shows the proportion of female wasps producing diapause offspring in the dark at a constant temperature of 18°C. After Saunders, D.S. (1973), *Science,* **181**, 358–360, Fig. 1, American Association for the Advancement of Science, Washington.

Night-length measurement in the aphid *Megoura viciae* is similarly achieved by a dark-period "hour-glass" or "interval timer" (Lees, 1968, 1973) although in this case, extension of the night by increasing T, or pulsing an extended night with supplementary light breaks as in a night-interruption experiment, failed to exhibit any circadian periodicity. Light-break experiments in cycles of $T=24$, however, produce the two "peaks" of light-sensitivity or long-day effect noted in other insects. Lees (1968) regards the dark-period "hour-glass" in *M. viciae* as a linked sequence of four reactions based on these responses. The second peak of light sensitivity (between the 5th hour of the night and the end of the critical night, 9.75 hours) is the third of these reactions, and represents the hour of the night which dawn must illuminate in order to produce the long-day response (wingless, parthenogenetic virginoparae). The short-day or long-night response (winged sexuales which mate and produce diapausing eggs) occurs when this third stage of the night is not illuminated. The similarity between the "hour-glass" clock in *M. viciae* and that in *S. argyrostoma* is clear: whether these forms of the photoperiodic clock reflect evolutionary convergence, or divergence from a common stock, is a matter for speculation. The clock in *M. viciae*, for example, might be a "defunct" oscillator, now only showing its "hour-glass" properties. Conversely, the similarities between the various species might be merely superficial.

The photoperiodic counter

Some plants, such as the cocklebur *Xanthium strumarium* and the duckweed *Lemna purpusilla,* will respond to a single photoperiodic cycle. The short-day plant *Pharbitis nil* will also produce its maximum flowering response following exposure to a single long night. In other species a *number* of cycles may be required: in *Glycine max* and *Kalanchoë blossfeldiana,* for example, the degree of the response increases over a number of days until it is saturated. In the birds, photoperiodic control of testicular and ovarian growth also seems to require a number of successive photoperiods of inductive length before induction can occur, or before it can be completed.

Similar phenomena can be observed in the insects. Some species, such as the larva of the phantom midge *Chaoborus americanus* may respond to a single diapause-terminating long-day cycle, although the majority of species require a sequence. In *Dendrolimus pini,* larval diapause can occur in any instar, but only after 30 to 35 short-day cycles have been experienced. More usually, however, diapause supervenes at a species-specific instar, but the *incidence* of diapause in the population is a function of the number of inductive cycles experienced during some prior "sensitive period". It is clear, therefore, that the programming of the central nervous system for subsequent development or diapause involves not only the measurement of night-length or daylength by the photoperiodic clock, but also the *summation* of photoperiodic cycles, either long or short, to a point at which induction can occur. The summation of light cycles is therefore an integral part of the photoperiodic response, and the observation that the number of cycles required is temperature-compensated (Saunders, 1971; Goryshin and Tyshchenko, 1970) has led to the concept of a *photoperiodic counter*. In insects, the interaction between the photoperiodic counter and the rate of development, and hence with a variety of environmental factors such as temperature and nutrition, has provided an empirical explanation for many aspects of photoperiodism. Some of these will be examined here.

In the parasitic wasp *Nasonia vitripennis*, the stage of the life cycle which is sensitive to photoperiod is the adult (female) instar; the diapausing stage is the mature larva of the next generation. Since the immature stages are insensitive to photoperiod, it follows that the *sensitive period* effectively comes to an end at the time of oviposition, the progeny being determined for either diapause or non-diapause post-embryonic growth whilst they are eggs within the maternal ovary. Eggs are deposited on practically every day of adult life, consequently the pattern of oviposition, and the type of progeny produced (diapause or non-diapause) from each egg batch provides a daily monitor of the physiological state of the mother. At short daylength ($<15\frac{1}{4}$ h/24) each female was found to produce developing progeny for the first few days, but then to switch abrubtly to the production of diapausing larvae. The "switch-over" occurred between the 6th and the 11th day in different individuals, but the mean number of such short-day cycles (or the *required day number* RDN) was the same whether the photoperiod was 6 hours per day or 14. At the critical daylength, however, the RDN became abruptly greater, until at long daylength (>16h/24) over 20 such cycles were required before any diapause was observed (figure 5.11). These changes in the RDN accounted for the effects of photoperiod on the production of diapause larvae (Saunders, 1974).

It was also found that the rate of oviposition ($=$ the sensitive period SP) was temperature-dependent, but the required day number was not: at temperatures between 30° and 15°C, for example, the RDN showed a Q_{10} of 1.04. Consequently, at high temperature (30°) most of the eggs were deposited before the wasps were programmed for diapause production, and the number of diapause larvae in the progeny was low (27 per cent). At

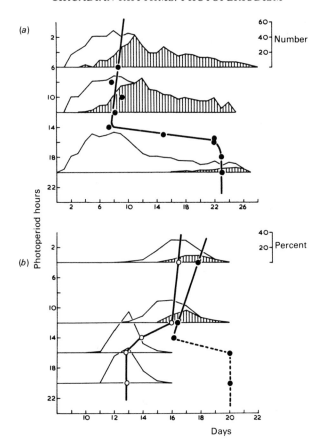

Figure 5.11   The interaction between the sensitive period (SP) and the required day number (RDN) in (*a*) *Nasonia vitripennis* and (*b*) *Sarcophaga argyrostoma.*

In (*a*) the whole of the adult life constitutes the SP and is represented here by the oviposition curve; it is essentially the same at all daylengths. The number of photoperiodic cycles required to raise the proportion of diapause to 50 per cent (the RDN), however, is low at short daylength, and becomes abruptly greater as the critical daylength is passed. Consequently at short daylength (<15/24) the proportion of diapausing offspring is high, whereas at long daylength (>15/24) it is low.

In (*b*) the period of larval development constitutes SP. At short daylength it is protracted, but at long daylength it is significantly shorter. The RDN at short daylength, however, is small so that a high proportion of the larvae enter diapause in the pupal instar. At long daylength the RDN is presumed to be high, and none of the larvae become dormant.

Open circles—SP; closed circles—RDN. The polygons show the proportion of the eggs produced (*a*) or puparia formed (*b*) per day; the shaded portions represent those insects entering diapause.From Saunders, D.S. (1975), *Insect Clocks,* Fig. 8–4, Pergamon Press, Oxford.

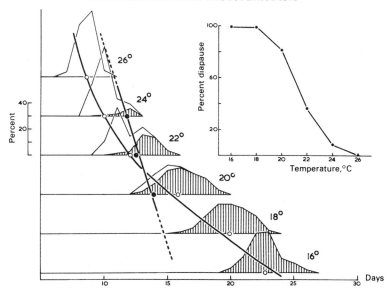

Figure 5.12 The effect of temperature on the induction of pupal diapause in *Sarcophaga argyrostoma* at short daylength (*LD 10*:14), showing the interaction between the temperature-dependent sensitive period (SP) and the temperature-compensated required day number (RDN). The polygons show the proportion of each batch of larvae forming puparia each day; the shaded portion of the polygons those larvae which became diapausing pupae.
Inset: The effect of temperature on the proportion of diapause pupae at *LD 10*:14. See text for details. After Saunders, D.S. (1975), *Insect Clocks,* Fig. 8–7 Pergamon Press.

low temperature (15°), on the other hand, the converse was true (91 per cent). Depriving the wasps of host puparia was also found to delay oviposition, but not the RDN, and therefore to have a predictable effect on the incidence of diapause. Other factors, such as geographical strain or host species, however, were found to affect the required day number rather than the sensitive period.

Diapause induction in the flesh-fly *Sarcophaga argyrostoma* is similarly a result of an interaction between the length of the sensitive period (larval development) and the required day number (Saunders, 1971), although in this species the *larval* sensitive period effectively comes to an end at puparium formation and the diapause stage is the pupa. The required day number (measured as the number of photoperiodic cycles required to raise the percentage of diapause in a day's batch of pupae to 50 per cent) was found to be about 13 to 14 at short daylength (*LD 10*:14), and to be temperature-compensated, whereas the duration of the sensitive period (larval development) was not (figure 5.12). The sensitive period was also shortened by starvation or premature extraction of the larvae from their food, in which case the larvae round up to form puparia earlier than in the controls, or lengthened by allowing the mature larvae to wander in wet sawdust or in constant light, both of which delay puparium formation. It was found that any factor which shortened the length of the sensitive period lowered the incidence of pupal diapause, and vice versa. In *S. argyrostoma* the sensitive period is also affected by photoperiod, being shortened at long days and lengthened at short days, even at the same temperature. The interaction between the sensitive period and the required day number is therefore synergistic, inductive short daylengths being experienced more often (figure 5.11).

In *S. argyrostoma* the larvae entered the pupal diapause provided that a sufficient number (13–14) of short-day cycles (<14 h$/24$) has been experienced before puparium formation terminated the sensitive period. The events leading up to and involved in puparium formation, such as the release of the hormone ecdysone, are "programmed" during the first 24 hours following the moult to the third and final larval instar (Zdarek and Slama, 1972). It is interesting, therefore, to note that, at very low temperature (8 to 12° C), the required day number has been accumulated by this time, and the larvae fail to pupate, as though they were in a *larval* diapause. Such larvae, however, although clearly "healthy" when they leave the meat, do not survive to puparium formation. Larvae at the same low temperature, but at *long* daylength (*LD 16*:8), on the other hand, do survive to pupation.

A third insect species in which the summation of photoperiodic cycles is temperature-compensated is the moth *Acronycta rumicis* (Goryshin and Tyshchenko, 1970). At temperatures between 18° and 26°C the length of the larval development was dependent on temperature, but the number of either long-day (*LD 22*:2) or short-day (*LD 12*:12) cycles applied at the end of larval development and required to produce a "critical" (50 per cent) level of diapause, was roughly the same at all temperatures tested. It is probable that such mechanisms are widespread, if not ubiquitous in insects, and perhaps other organisms. In summary, it is clear that photoperiodic induction is a two-stage process:

1. the measurement of a particular daylength or night-length (either long or short) by the photoperiodic clock,
2. the summation of a critical number of such cycles by the photoperiodic counter.

Both are aspects of circadian rhythmicity, the first a question of the particular phase relationship adopted by the oscillator(s) to the light cycle in steady-state entrainment; the second represents a day-by-day summation of such information through successive entrained circadian cycles. It is clear, therefore, that any attempt to analyse the photoperiodic response must take into account both of these mechanisms.

The spectral sensitivity and intensity threshold for the photoperiodic response

The question of the intensity threshold is complicated by the fact that insects may be in "cryptic" situations or, since the brain contains the photoperiodic receptors (Chapter 8), much of the light energy may be absorbed by the integument and other tissues overlying the brain. For instance, the larvae of the oriental fruit moth *Grapholitha molesta* show a threshold of about 10-30 μW cm^{-2} whilst burrowing in young apples, but the light actually reaching the brain must be much less than this. Conversely, the larvae of the pitcher plant midge *Metriocnemus knabi* respond to less than 0.025 μW cm^{-2} but normally live concealed in the water that collects within the plant. Williams *et al.* (1965) have claimed that although the pupa of the silkmoth *Antheraea pernyi* is enclosed in a very opaque cocoon, the geometry of this cocoon makes it an ideal "light-integrating sphere" which collects within it light as a blue haze. About 1 per cent of the incident light collects within the cocoon, and the brain is saturated by about 10μW cm^{-2} of this blue light.

In many insects the photoperiodic threshold is below the intensity of the full Moon, and this raises the question of the potential role of moon-light as a "night interruption" which might reverse the short-day (long-night) induction of diapause during the autumn months. Interference by moon-light is considered unlikely, however, particularly because a number of coincidences between ϕ_i and light are required to reverse the diapause response and—because of the period difference between the lunar day (24.8 hours) and the solar day (24.0 hours)—a sufficient number of lunar interruptions is unlikely to occur in the latter half of the night before the period of brightest moonlight encroaches on sunrise and no longer functions as a perturbation.

The spectral sensitivity of the photoperiodic response is very variable. Some species show a maximum response in the blue-green region of the spectrum and an almost complete insensitivity to red (Lees, 1971); these action spectra are similar to those for phase-shifts of the circadian system (Chapter 3). In others, such as *N. vitripennis,* sensitivity extends well into the red end of the spectrum (Saunders, 1975*b*). Evidently the pigment involved in the photoperiodic chromophore is equally variable.

CHAPTER SIX

BIOLOGICAL RHYTHMS
WITH AN ANNUAL PERIODICITY

SEASONALLY-APPROPRIATE CHANGES IN PHYSIOLOGY OR BEHAVIOUR can be synchronized to an annual cycle by at least two methods. Many organisms, both plant and animal, respond to day- or night-length with a photoperiodic "clock", which may or may not be a function of the *circadian* system (Chapter 5). Other species, particularly long-lived plants and animals, may use an endogenous "calendar" possessing the properties of an inherent *circannual* (or circannian) rhythm with a natural period (τ) close to a year. The literature on seasonal cycles in higher plants and animals abounds with evidence for such cycles, and their occurrence has been known or suspected ever since temperate species were transported to the tropics and observed to persist with their seasonal cycles of leaf unfolding, flowering, or breeding, despite the almost total lack of seasonal clues.

It is possible that "classical" photoperiodism and circannual cycles of responsiveness to daylength coexist, at least in long-lived animals such as birds and mammals. It is known, for example, that testicular growth in birds is under photoperiodic control, but that the birds may pass through periods when such response may occur (i.e. in the spring) and other (refractory) periods when photoperiod has little or no effect (i.e. in the autumn). Perhaps all cases in which long-lived organisms respond to increasing or decreasing daylengths, or to sequences of daylengths (short to long, or long to short), should be considered as containing a circannual element.

If endogenous circannual rhythms are analogous to those rhythms with a daily (circadian) periodicity, they ought to possess the following properties:

1. A *"free-running"* period (τ) which, in the absence of a Zeitgeber, is *approximately* equal to the period of the sidereal year (365 days), although it may vary from 40 to 60 weeks.
2. A *temperature-compensated* period for the free-running (unentrained) circannual oscillation.
3. *Entrainment* in natural conditions to a Zeitgeber, whatever that might be, so that the period of the oscillation (τ) becomes exactly 12 months.

100

4. A differential sensitivity of the circannual oscillation to the phase-resetting effects of the Zeitgeber, according to the *phase* of the oscillation so perturbed. This supposes that a *phase-response curve* should be available.

It has been shown or suggested that circannual rhythms underlie such seasonal responses as testis growth, body weight changes, moulting and migratory restlessness in birds; locomotor activity rhythms in lizards; moulting and reproduction in crayfish; hibernation/activity cycles, body weight, pelt changes, oestrous cycles and antler cycles in mammals; and annual cycles of diapause and development in long-lived or semivoltine insects. In this chapter we will examine some of the evidence suggesting that these seasonal cycles possess the properties of circannual rhythms. We will also see that some of these rhythms are equivalent to the circadian *behavioural rhythms* which govern daily locomotor activity in an individual organism, whereas at least one example (the "carpet" beetle *Anthrenus verbasci*) is the almost exact counterpart of a "once-in-a-lifetime" *developmental* or "gated" oscillation like that controlling the rhythm of pupal eclosion in *Drosophila pseudoobscura* (Chapter 2).

Circannual rhythms in arthropods

Larvae of the "carpet" beetle *Anthrenus verbasci* feed on material of animal origin and are commonly found in old house sparrows' nests. In their natural environment, the life cycle occupies at least two years (i.e. the species is semivoltine), the first winter being spent as a young larva in diapause, the second winter, again in diapause, as a mature larva. After the second winter the reactivated larva pupates, and the adult emerges in the following spring. Although most individuals complete their development in two years, a small proportion may take three or more.

Blake (1959) found that populations of *A. verbasci* maintained in the laboratory at constant temperature, constant humidity and in continuous darkness (DD), showed a persistent rhythm of roughly "annual" periodicity in which periods of development were followed by periods of diapause. This rhythm produced "pulses" of pupation and adult emergence roughly corresponding to those individuals completing development after one, two or three years, respectively (figure 6.1). The interval between these peaks, however, was found to be 41–44 weeks, rather than the 52 weeks observed in outdoor populations. This interval was recognized by Blake as the "basic periodicity" and represents the free-running period (τ) of a circannual rhythm. Since the free-running period was considerably less than a year, the laboratory populations rapidly lost synchrony with the natural seasons.

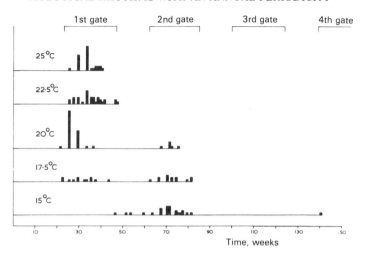

Figure 6.1 The circannual rhythm of pupation in mixed-age populations of the beetle *Anthrenus verbasci*. Frequency of pupation times when larval development has occurred in constant conditions of temperature, humidity and darkness. A black square represents the time of pupation, to the nearest week, of an individual. Note that the larvae at higher temperatures are able to utilize the first gate; at lower temperatures, however, an increasing proportion of them are required to wait until the next. After Blake, G. (1959), *Nature*, **183**, 126–127, Macmillan.

The period of the rhythm in *A. verbasci* was found to be the same in constant (i.e. unchanging) photoperiod; it was also temperature-compensated. The temperature, however, controlled the *proportion* of the population utilizing each annual "gate". At 25° and 22.5° C, for example, all of the larvae completed development in one cycle and emerged in the first peak. At 15° C, on the other hand, all of the individuals went through two cycles of development and diapause, and emerged in the second peak. At intermediate temperatures (20° and 17.5° C) some of the larvae, presumably the faster developers, were able to utilize the first gate, whereas the slower members of the population were forced to undergo two annual cycles and emerge in the second, about 41 weeks after the first. At least one beetle at 15° C seemed to miss even the third gate, and eventually emerged 140 weeks (or "four years") after the start of the experiment. The similarities between this rhythm which gates developmental events in a mixed-age population by a circannual rhythm, and that governing daily eclosion peaks in *D. pseudoobscura* (Chapter 2), are striking.

In later papers it was demonstrated that the Zeitgeber responsible for the entrainment of the free-running circannual rhythm ($\tau = 41$–44 weeks) to the exact period of the sidereal year (52 weeks; 365 days) was the annual *change* in daylength and, to a lesser extent, the annual change in temperature. *Decreasing* daylength, for example, delayed pupation by 13 weeks (from October to January), thus achieving synchrony with the natural seasonal cycle.

A circannual rhythm is also known to regulate the annual rhythms of moulting and reproduction in the cave crayfish *Orconectes pellucidus* (Jegla and Poulson, 1970). This permanently cave-dwelling species lives in an environment in which light is completely absent and the annual temperature range is small. However, in this almost seasonless environment, its cycle of reproduction is synchronized to the time of the year when the surface run-off is highest, and the water entering the cave system brings in the annual flux of organic matter which forms the basis of the food chain in the cave environment. Reproduction is maximized at these times so that the addition of young crayfish into the population occurs when the food supply is most plentiful.

When these crayfish were maintained in the even more seasonless environment of the laboratory (DD and 13°C) the annual cycles of moulting and reproduction were observed to persist, but with a considerable lack of synchrony. Individual crayfish showed circannual rhythms with periods varying between 338 and 396 days, and thus rapidly became out of phase both with other members of the population and with the calendar year. The fact that synchrony is achieved in their natural environment points to the existence of a Zeitgeber. Jegla and Poulson considered that this function is provided by subtle changes in the volume of water entering the cave system, its rate of flow, or perhaps its temperature.

Circannual rhythms are almost certainly more widespread among the Arthropods than these two examples might suggest: these two merely represent examples in which the essential properties of an endogenous circannual rhythm, as outlined in the opening paragraphs, have been unequivocally demonstrated.

Circannual rhythms in birds

The early literature on seasonal cycles in birds contains numerous references to inherent annual cycles, but only within the last decade have birds been kept for protracted periods in unchanging conditions and the truly circannual properties of these rhythms demonstrated beyond doubt. Circannual rhythms are now known to be involved in the cycles of breeding, moulting and migratory restlessness (Zugunruhe) in a number of

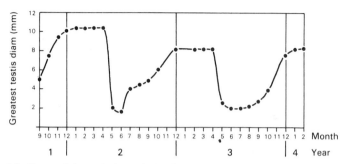

Figure 6.2   Seasonal change in the testicular size of the weaver-finch *Quelea quelea* kept in an unchanging photoperiod (*LD 12*:12). After Lofts, B. (1964), *Nature*, **201**, 523.

species. They are particularly important in those birds which inhabit tropical areas but synchronize their breeding activity to a particular season. They are also important in those *migrants* which winter in the tropics where reliable seasonal cues (i.e. photoperiodic changes) are absent or unreliable, and in trans-equatorial migrants which are exposed to complicated sequences of daylength changes as they cross and re-cross the equator.

Benoit and his associates maintained domestic drakes in constant light or in constant darkness, and demonstrated the persistence of long-period cycles of testicular growth and regression. These cycles deviated so far from 12 months, however, that it is doubtful whether they should be regarded as "circannual". Lofts (1964), on the other hand, demonstrated the persistence of a cycle of testicular growth in the tropical weaver-finch *Quelea quelea,* when maintained in an unchanging *LD 12*:12 light regime for over two years; in this case the periodicity was so close to 12 months that it is possible that a subtle seasonal change in the environment was unaccounted for (figure 6.2). However, most of the recognized and expected properties of circannual rhythms, as outlined in the opening paragraphs of this chapter, have now been demonstrated in recent work with migratory warblers (Gwinner, 1971) and with the common starling (Schwab, 1971).

Gwinner (1971) demonstrated that circannual rhythms participate in the timing of Zugunruhe, body weight and moult cycles of the willow warbler *Phylloscopus trochilus,* the wood warbler *P. sibilatrix,* and in the chiffchaff *P. collybita.* These are Palaearctic species which winter in central or southern Africa. Those wintering in central Africa do so in an area which is deficient in reliable seasonal cues, such as those afforded by changes in photoperiod at higher latitudes. Those wintering further south cross from areas with decreasing photoperiods to areas with increasing photoperiods, or vice versa, each time they undergo migration. An endogenous "calendar" is therefore invaluable for the timing of the winter moult and for the initiation of the spring (northerly) migration.

In the laboratory, birds were kept for up to 27 months in LD *12*:12. Nevertheless, the annual cycles of Zugunruhe, moult and, to some extent, body weight, were found to persist, but also to deviate from an *exact* 12-month periodicity, a result which is consistent with the conclusion that circannual rhythms are involved. Gwinner also demonstrated a differential sensitivity to the phase-resetting effects of photoperiod, according to the phase of the circannual oscillation. Birds transferred from natural lighting conditions to *LD 12*:12 in spring and early summer showed a phase advance (+*Δø*) of about 10 days in the ending of the postjuvenile moult; those transferred to *LD 12*:12 in late September, however, showed a phase delay (–*Δø*) of 20–40 days in the beginning of their postnuptial moult. Conversely, birds transferred to long daylength (*LD 18*:6) in the spring showed a phase delay of about 10 days, but a phase advance of 20–40 days when transferred in the autumn. These observations, similar to those in *Anthrenus verbasci,* provide the basis for a circannual phase response curve, and amply demonstrate the similarities between circannual and circadian rhythms, and the essentially oscillatory nature of them both.

The annual testicular cycle of the starling *Sturnus vulgaris* has also been studied in a constant photoperiodic environment (Schwab, 1971). At *LD 12*:12 and 22–28° C, testis growth began in December and reached average

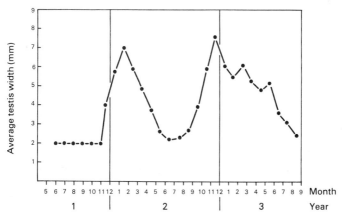

Figure 6.3   Seasonal change in testicular size in the European starling *Sturnus vulgaris* in unchanging photoperiod. In *LD 12*:12 a free-running circannual rhythm in testicular periodicity is observed. After Schwab, R.G. (1971), in *Biochronometry,* ed. M. Menaker, pp. 428–447, Fig. 3, National Academy of Sciences, Washington.

spermatogenesis by mid-January. The testes became quiescent again by June, but commenced a second testicular cycle, achieving spermatogenesis by the following October or November (figure 6.3). The free-running circannual cycle was therefore shorter (~9½ months) than the naturally entrained period (12 months), and the testicular cycle of captive birds soon drifted out of synchrony with the calendar year. Similar endogenous cycles were not observed, however, when the birds were maintained at LD *11*:13 or at LD *13*:11. In the birds' natural environment (38°N) the annual testicular cycle can be divided into four phases:

1. The interval between the end of testicular quiescence to spermatogenesis, a period which corresponds to the lengthening days between the winter solstice (*LD 9*:15) and the vernal equinox (*LD 12*:12); the rate of testicular growth during this period is a function of photoperiod.
2. Continuing testicular growth to spermatogenesis and subsequent involution of the testes to a quiescent state; this stage occurs between the vernal equinox and the summer solstice (*LD 15*:9).
3. The first half of the quiescent period between the summer solstice and the autumnal equinox; during this time it is impossible to stimulate precocious testicular growth by increasing the daylength (i.e. the "refractory" period).
4. The second half of the quiescent period, from the autumnal equinox to the winter solstice, characterized by an increasing sensitivity to photoperiod as midwinter approaches.

This sequence suggests that the response to photoperiod itself is governed by a circannual rhythm.

Circannual rhythms in mammals

Endogenous annual rhythms are known or thought to control such seasonal events as breeding, wool growth, antler development, and the behavioural and physiological changes associated with the hibernation/activity cycle in hedgehogs and a variety of rodent species. Early observations with domestic sheep indicated that the annual cycle of wool production, naturally highest in the summer months, persisted when the animals were maintained in continuous light (*LL*).In ferrets the annual oestrous cycle also persisted in *LL*, but became increasingly irregular and unsynchronized. In blinded ferrets the oestrous cycle occurred with an increasing frequency.

Most attention has been paid within recent years to those rodents which exhibit spectacular cycles of hibernation and activity, often "disappearing" and "appearing" with a seemingly mathematical accuracy at precise times of the year. The truly endogenous and circannual nature of these rhythms has now become clear. In ground squirrels of the genus *Spermophilus* (*Citellus*), Pengelley and Kelly (1966) and Pengelley and Asmundson

(1969) have shown that the rhythms of hibernation and the associated changes in body weight and temperature persist in constant laboratory conditions. In *S. lateralis* at LD *12*:12, for example, the period τ departed significantly from 365 days when in its unentrained ("free-running") condition, but was not affected by temperature (12° or 3°C). The rhythm of body weight, which reached its maximum at the onset of hibernation and its minimum at the spring arousal, was not affected by the availability of the food. There were also no differences between the sexes, between adults and juveniles, or between castrated males and intact controls. Out of a total of 61 circannual periods measured, only 13 were longer than 365 days; the longest was 445 days, the shortest 229.

Heller and Poulson (1970) have also recorded observations on circannual rhythms in ground squirrels (*Spermophilus* spp.) and chipmunks (*Eutamias* spp.) kept continuously at LD *12*:12 and at 16° or 5°C. They observed rhythms of body weight, hibernation, reproductive condition and daily water consumption. The free-running period (τ), measured from one "terminal arousal" (from hibernation) to the next, varied from 44 to 59 weeks with the majority being less than a year.

Goss (1969) described the endogenous circannual rhythm regulating cycles of antler growth in the deer *Cervus nippon*. Unlike the hibernating rodents and the starling, deer kept at an unchanging photoperiod of LD *12*:12 failed to replace their antlers, but those kept at LD *8*:16, LD *16*:8 or in *LL* showed persistent antler cycles with a periodicity close to 85 per cent of the sidereal year. The average τ-values were 287.3 days at LD *8*:16, 317.0 days at LD *16*:8, and 314.4 days in *LL*. Under natural daylight conditions, however, the rhythms became entrained to an exact 365 days. Evidence was adduced to suggest that the Zeitgeber responsible for this entrainment was the natural seasonal change in photoperiod.

CHAPTER SEVEN

BIOLOGICAL RHYTHMS WITH A TIDAL
OR LUNAR PERIODICITY

ANIMALS AND PLANTS LIVING IN THE INTERTIDAL ZONE ARE SUBJECTED to alternating periods of inundation and exposure. These tidal cycles which—on most coasts—recur twice every lunar day (24.8 hours), bring with them associated fluctuations in temperature, pressure, mechanical agitation due to the pounding of the waves, food supply, and possibly of light intensity and salinity. Some organisms may be more active when covered by the water, others more active when it recedes, the duration and onset of such activity depending on the vertical position of the organism on the beach. Local tidal conditions are also very variable, not only in tidal height, but in the relative height of the two tides per lunar day (the semi-diurnal inequality), and in the timing of the tides along a particular stretch of coast as the tidal stream advances. Most of these variations are reflected in the activity cycles of intertidal creatures. Other species, inhabiting the highest or lowest reaches of the tidal cycle, are only covered or exposed by the highest or lowest of the spring tides which occur every 14.7 days, or one-half of a lunar cycle (29.5 days).

Many of the organisms from these environments show *persistent* biological rhythms in a wide variety of behavioural or physiological function, even when removed to the uniform and tideless environment in the laboratory. Those from the mid-tidal range may show persistent *circatidal* rhythms with an endogenous period close to 12.4 hours. Those from the extreme ranges of the spring tides may show a *semi-lunar* or *circasyzygic* rhythm with a period of 14.7 days. This chapter will look at some of these rhythmic systems, examine their properties, and see how they compare with the better-known circadian system. Also included will be those rhythms of a truly *lunar* periodicity, which are to be seen in some marine, fresh-water, or even terrestrial organisms.

One subject of particular interest is the nature of the tidal periodicity. The persistence of an apparently endogenous rhythm with an interval between activity peaks of approximately 12.4 hours does not necessarily mean that the tidal rhythm is regulated by a truly circatidal oscillation with a period close to 12.4 hours. Alternative explanations might include a single *circadian* rhythm with a period close to 24.8 hours and a *bimodal* activity pattern, or perhaps *two* similar circadian rhythms held in a suitable out-of-phase relationship. "Circatidal" and circadian rhythms, however, may also coexist in tidal organisms, and lastly, some authors have attempted to explain the nature of some of the longer period rhythms by postulating a "combination" of higher-frequency oscillations, interacting on the "beat" principle.

"Circatidal" rhythms

Biological rhythms with a tidal periodicity and persisting in the tideless or otherwise constant conditions obtained in the laboratory, occur in a wide variety of intertidal plants and animals and concern an equally wide variety of behavioural and physiological events. These include the vertical migration rhythms of sand-dwelling flatworms, unicellular algae, diatoms and polychaetes; the rhythms of expansion and contraction in sea anemones; opening and closing of the shells, and filtration rates in bivalve molluscs; locomotor activity, oxygen consumption and colour change in a variety of crustacea; and swimming activity in fish. In most areas with a semi-diurnal equal, or nearly equal, tidal cycle these rhythms show two peaks of activity per lunar day. As with circadian and circannual rhythms, many of these tidal periodicities free-run with a period which

1. departs from the strictly tidal intervals observed in nature,
2. is temperature-compensated,
3. is sensitive to the entraining effects of a synchronizing agent or Zeitgeber associated with the tidal cycle.

A few examples of tidal rhythms will be described here in greater detail to illustrate some of these properties.

One of the first persistent tidal rhythms to be described was that in *Convoluta roscoffensis,* a flatworm occurring in certain localities on the coast of Brittany. In 1903 Gamble and Keeble reported that these worms migrate to the surface of the sand during the *day-time* low tides, but disappear into the sand as the tide rises, and also at night. The selective advantage of this behaviour seems to be two-fold: coming to the surface at low tide and in the light facilitates photosynthetic activity in the green algae the worms harbour within their tissues, and burrowing at high tide presumably prevents the worms being dislodged from the most favourable site on the beach. This tidal rhythm of vertical migration persists for 4–7 days in the laboratory in an approximate synchrony with the tidal cycle on the beach from which the worms were collected. As in their natural habitat, however, the worms do not surface in the dark; consequently the rhythm persisted in *LL* but not in DD.

Rao (1954) described a rhythm in the filtering rate of the mussel *Mytilus* spp. which persisted in the laboratory for over four weeks, whether in DD, *LL* or in a natural *LD* cycle, roughly in phase with the tides of their natural habitat. The period of this rhythm was temperature-compensated (between 9° and 20°C) and considered to be endogenous. Good evidence for this endogeneity was obtained in a translocation experiment in which mussels were collected from a beach near Cape Cod, Massachusetts, and rapidly transported to California. When they were subsequently maintained in laboratory conditions, the rhythm persisted with a Cape Cod tidal cycle, but when the mussels were exposed to local tidal influences they became re-entrained to the local (Californian) tidal cycle within about a week. The similarities between this experiment and that described for honey bees (Chapter 4) are clear. Unfortunately, however, subsequent and extensive attempts to confirm the existence of persistent rhythms in mussels, either tidal or daily, have not met with success (see Enright, 1963); although these observations do not negate those of Rao, his apparently clear results become equivocal.

Rhythmic swimming activity in the littoral fish *Blennius pholis* was found to persist for at least 5 days in the absence of a tidal cycle, and in constant conditions of light and temperature (Gibson, 1965). In this free-running condition the period of the rhythm became somewhat *longer* than the semidiurnal tidal interval (12.4 hours): the mean value in DD was 12.56 hours, in *LL* 12.5 hours. Both the persistence of the rhythm and its departure from the exact periodicity of the tidal environment constitute evidence for the endogenous nature of the rhythm. More persuasive evidence for this conclusion was obtained by Naylor (1963) who showed that a single non-recurrent low-temperature pulse (4°C for 15 hours) was sufficient to re-initiate a tidal rhythmicity in the crab *Carcinus maenas* after it had "faded out" in the laboratory, or in specimens from a non-tidal dock which showed no tidal component in their activity rhythms. Williams and Naylor subsequently showed that crabs raised in the laboratory from the egg showed only a circadian rhythmicity with maximum activity during their subjective night, until a period of chilling initiated outbursts of activity at 12.4-hour tidal intervals. Since these laboratory-reared crabs had never been exposed to tidal cycles, and the single low-temperature pulse conveyed no "information" about tidal periodicity, the tidal rhythm in *C. maenas* must be endogenous.

Apart from the equivocal example of *Mytilus edulis* described above, the best evidence for temperature compensation of the tidal period also comes from the work of Naylor (1963) on *Carcinus maenas*. Freshly-collected crabs were placed in dim constant light and at constant temperatures of 10, 15, 20 and 25°C; during a period of three days there was no obvious lengthening or shortening of the intervals between activity peaks.

Coexistence of circadian and "circatidal" rhythms in littoral organisms

Many authors have demonstrated or postulated the existence of both circadian and "circatidal" components in the same organism, the former being synchronized in nature by the light cycle, and the latter by some component of the tides. In some cases the two rhythmic systems appear to interact to produce a complex pattern of behaviour.

In the green shore crab *Carcinus maenas,* for example, maximum locomotor activity occurs during the daytime high tides and at night (Naylor, 1958). When maintained in a laboratory aquarium and tested at intervals in an actograph, crabs displayed a prominent tidal rhythm for several weeks, after which activity became circadian with pronounced nocturnal peaks. Crabs collected from a non-tidal environment (a floating dock), and specimens of a related species *C. mediterraneus* from a Naples shoreline with a low tidal amplitude, showed only the circadian component, or a very weak tidal rhythmicity. In fiddler crabs of the genus *Uca,* "circatidal" rhythms of locomotor activity may coexist with circadian rhythms of colour change (see Fingerman, 1960) and of oxygen consumption. Both components have also been demonstrated in laboratory-held specimens of *Sesarma reticulatum* and *Emerita asiatica*. In the latter, the interaction between the tidal and circadian components results in the nocturnal high-tide peak of activity being greater than that during the day.

Of particular interest in this connection are the vertical migration rhythms in sand-dwelling organisms such as that described above for *Convoluta roscoffensis,* in which their appearance at the surface occurs during the daytime low tides, but not at night. Similar rhythms have been described for the diatom *Hantzschia virgata* (Palmer and Round, 1967). These organisms also appear on the surface of the sand during *daytime* low tides, often in numbers large enough to colour the sand a golden-brown. With the flood tide, however, the diatoms burrow into the sand. The vertical migration is inhibited by continuous darkness, but persists in constant light and in the absence of tidal stimulation. Palmer and Round

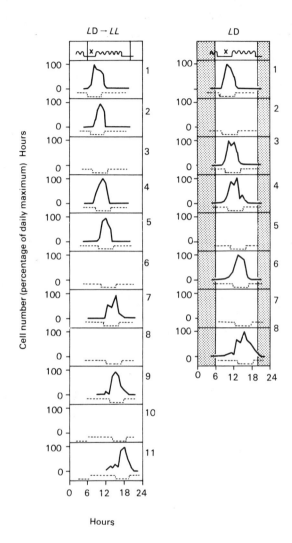

Figure 7.1 Persistence of the vertical migration rhythm of the diatom *Hantzschia virgata* in the laboratory, and in constant light (*LL*) or a light cycle (*LD*). Consecutive days run from top to bottom. The times of low water in their natural habitat are indicated by the dotted line below each record. x — time of collection. Note that the peaks of surface activity occur about 50 minutes later each day, the rhythm of activity having a period of about 24.8 hours. After Palmer, J.D. and Round, F.E. (1967), *Biol. Bull, Woods Hole,* **132**, 44–55, Fig. 2.

observed that the peaks of surface activity followed a "circatidal" rhythm, appearing about 50 minutes later each day until the low-tide peak approached sunset, when the rhythm was rephased to the early morning hours of light (figure 7.1). This behaviour was interpreted as an interaction between a tidal rhythm theoretically capable of producing two peaks of surface activity every double-tidal or lunar-day cycle (24.8 hours), and a solar-day (circadian) rhythm responsible for the night-time suppression of one of the tidal peaks. The selective advantage of this behaviour, however, was far from clear, since sufficient light apparently penetrates the sand to obviate the need for periodic surfacing to maximize photosynthesis.

The "nature" of the tidal rhythmicity: one oscillator or two?

It was pointed out in the opening paragraphs of this chapter that the demonstration of persistent peaks of activity recurring at approximately 12.4 hourly intervals does not necessarily mean that the organism possesses an endogenous *circatidal* rhythm with a natural period (τ) of that duration. Indeed, it now seems more likely that such periodicities are a reflection of either a *single* but bimodal circadian rhythm, or of *two* similar circadian rhythms held in antiphase. In both possibilities the circadian rhythms are entrained in nature by some component of the tidal cycle rather than by changes associated with the solar day. Evidence for this point of view has been obtained from the very precise tidal rhythms in certain sand-beach crustacea (*Synchelidium* and *Excirolana*) on the Pacific coast of North America.

Tidal cycles on the Californian coast near La Jolla are of the mixed semidiurnal type, alternating between a *diurnal* tide (one high water per 24.8 hours) and a *semidiurnal* form (two high waters per 24.8 hours). Over a period of several days, the tidal pattern changes from the diurnal type through semidiurnal with very *unequal* tidal amplitudes (e.g. alternate high and low tidal crests) to semidiurnal *equal* tides, and back again. When of the unequal semidiurnal type, the intervals between the two high tides may also depart significantly from 12.4 hours, although the sum of the two intervals is always approximately 24.8 hours.

The amphipod *Synchelidium* sp. lives in the "uprush zone" on fine sandy beaches and synchronizes its swimming activity to the phase of the high tide; at other times it burrows into the sand. This tidal periodicity is endogenous and persists in the laboratory in the absence of tidal stimulation (Enright, 1963). Moreover, most of the complexities in their tidal environment are reflected in the laboratory rhythm. Thus newly-collected

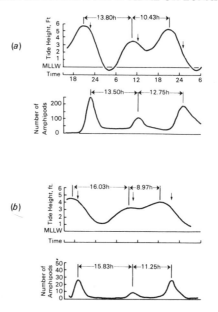

Figure 7.2 (*a*) (*b*)   Tidal rhythmicity in populations of freshly collected *Synchelidium* sp. (Crustacea, Amphipoda). The upper panel of each figure shows the heights and intervals between the two high tides in their natural habitat. The lower panels show the swimming rhythm in constant and tideless conditions in the laboratory. After Enright, J.T. (1963), *Z. vergl. Physiol.*, **46**, 276–313, Fig. 1, 2, Springer Verlag.

populations of *Synchelidium* reflect the semidiurnal inequality of the tidal cycle occurring on the beach during the few days immediately prior to collection. When, for example (figure 7.2*a*) the intervals between the two tidal crests were 13.80 and 10.43 hours (total cycle time 24.23 hours), the rhythm of swimming activity duplicated both the relative amplitude of the two tidal peaks *and* the intervals between them, the latter being 13.50 and 12.75 hours, respectively. A similar group of amphipods collected when the tidal intervals were even more dissimilar 16.03 and 8.97 hours) showed tidal peaks 15.83 and 11.25 hours apart when observed in the laboratory (figure 7.2*b*). The total cycle time (between first and third activity peaks) in constant conditions was longer than the natural tidal period. In 22 experiments, for example, the mean endogenous period was 26.48 hours (range 25.25–27.75 hours) whereas the intervals between corresponding tidal crests was 24.98 hours (24.53–25.65 hours). This departure from the natural tidal cycle in the absence of entrainment constitutes strong evidence of endogeneity.

The isopod *Excirolana chiltoni* also reflects the tidal pattern, both amplitude and interval, when it is transferred to the tideless conditions of the laboratory (Klapow, 1972; Enright, 1972). Figure 7.3 shows, for example, that populations transferred to the laboratory from tides of the diurnal type persist with one peak of swimming activity every lunar day, whereas those transferred from equal or unequal semidiurnal tidal cycles persist with two activity peaks per lunar day, the latter with a marked

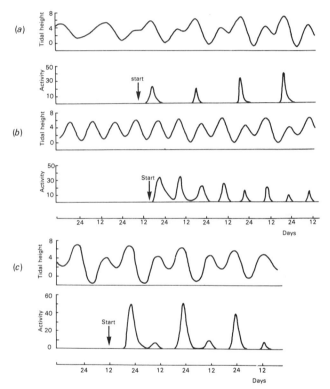

Figure 7.3 Tidal rhythmicity in populations of freshly collected *Excirolana chiltoni* (Crustacea, Isopoda). The upper panels show the heights and intervals between the high tides in their natural habitat. The lower panels show the swimming rhythm in the constant and tideless conditions in the laboratory.

(*a*) population collected after entrainment by a diurnal tidal cycle showing *one* peak of activity every 24.8 hours.

(*b*) population collected after entrainment by a semidiurnal equal tidal cycle; *two equal* peaks of activity per cycle.

(*c*) population collected after entrainment by a semidiurnal unequal tidal cycle.

Note how the rhythms of activity in the laboratory match both tidal height and interval. After Klapow, L.A. (1972), *J. comp. Physiol.,* **79**, 233–258, Figs. 1A 7 and B, 2, Springer Verlag.

alternation in amplitude. The period (τ) of the double-tidal interval in its free-running condition does not, however, differ significantly from the natural cycle, but the *timing* and the *form* of the laboratory rhythm deviate from the *concurrent* tidal cycle (figure 7.3). In other words the laboratory rhythm must be the result of events which occurred before collection; it is therefore endogenous and free-running.

The endogenous activity rhythms of *Synchelidium* sp. and *E. chiltoni* are difficult to explain if we postulate control by a single circatidal oscillator with a period τ close to 12.4 hours. This explanation, for example, would necessitate a complicated additional mechanism which adjusts (suppresses or amplifies) alternate peaks, and also adjusts the intervals between them. Enright (1963) proposed two possible alternatives:

1. A single circadian rhythm normally entrained by the tidal cycle to 24.8 hours, but with the potential for a bimodal pattern. Such bimodal patterns are known in circadian systems and include those in which the amplitude of the first or the primary peak determines the time interval to the next or following peak.
2. Two similar, but unimodal, circadian rhythms, held in 180° antiphase, and with some degree of mutual independence to allow for the changing phase relationships between alternate activity peaks.

Klapow (1972) pointed out that the "two oscillators" might occur in each individual in the population, or part of the population may swim in the first peak and other individuals in the second. When swimming animals were removed to a second container, however, they continued to show activity in both peaks, thereby demonstrating that *each animal* shows a semidiurnal periodicity. Single specimens of *E. chiltoni* also show precise semidiurnal tidal cycles of activity, sometimes persisting for months in constant non-tidal conditions (Enright, 1972).

The first alternative (a single bimodal rhythm) has received recent support from entrainment experiments with a "wave simulator" (Enright, 1972). Single two-hour stimuli similar to those from turbulent water or waves on the beach were applied to populations of *E. chiltoni* under free-running conditions, and at different phases of the double-tidal cycle. The phase response curve obtained was itself bimodal with an amplitude (advance or delay phase-shifts) of about two hours. Since the activity peaks were not separately entrainable, Enright considered that the simplest interpretation was that the rhythmic system (the "tidal clock") was a bimodal oscillator.

Entrainment of tidal rhythms

Most tidal rhythms persist with a periodicity close to 12.4 or 24.8 hours when kept in the absence of tides, but in LD *12*:12 or natural LD cycles. Therefore, apart from those examples (*Carcinus maenas, Hantzschia virgata,* etc.) in which a solar-day-entrained circadian component is present, true tidal rhythms are not entrained by the day-night cycle. The circadian rhythm or rhythms postulated to control tidal rhythmicity in *Synchelidium* sp. and *Excirolana chiltoni* thus differ from "normal" circadian rhythms in lacking this sensitivity to light. According to Palmer (1973) "rhythms with a period of 24.8 hours in nature, or thereabouts, when moved into constant conditions are unique, and not simply a part of the spectrum of circadian frequencies longer than 24 hours". He refers to such rhythms as "circalunadian". However, these periodicities are well within the circadian "spectrum" (see Chapter 2), and only differ in the manner of their coupling to the environment: the erection of a new term therefore seems unwarranted.

Biological rhythms, whether circadian, circannual or "circatidal" have an endogenous period (τ) which deviates in its free-running state from the environmental periodicity (T) which they have evolved to "match". In natural conditions, however, τ is corrected in each cycle so that $\tau = T$, in a process of entrainment to the natural cycle. This procedure requires and implies that a differential sensitivity exists towards a component of the environment (a Zeitgeber) which gives rise to adjustments in phase (either advance or delay) according to the phase of the cycle so perturbed. In tidal rhythmicity, candidate Zeitgebers might include cycles of temperature, chemical change, or mechanical stimuli associated with the tides. Direct entrainment by lunar influences is also possible, but in *Synchelidium* and *E. chiltoni,* with their complex adoption of a semidiurnal inequality, the differences in amplitude and interval are not directly coordinated with irregularities in the lunar zenith or nadir; this constitutes *a priori* evidence that the Zeitgeber is tidal rather than lunar.

Cycles of submergence and exposure, chemical change, availability of food, or of oxygen tension, have no entraining effects in *Excirolana* (Enright, 1965), and temperature cycles are generally regarded as erratic and unreliable. The most likely Zeitgebers, therefore, are hydrostatic pressure or mechanical agitation, such as that produced by the surge of wave action on the beach. Mechanical agitation has proved to be the most efficient synchronizer in *E. chiltoni,* and wave action can be reproduced in the laboratory using "wave simulators" (Enright, 1965). Klapow (1972) subjected samples of this isopod to 7 days of periodic agitation, consisting of 30 minutes of swirling water every afternoon, alternating with 120 minutes of the same in the morning, the two treatments being separated by

12.5 hours. This procedure entrained the activity rhythm to a semidiurnal periodicity with one large peak and one smaller peak. A single stimulus per day, on the other hand, moulded the behaviour of the isopods into a *diurnal* tidal rhythm. Enright has since found that single 2-hour stimuli of simulated wave action are sufficient to cause either phase delay or phase advance in the rhythm of *E. chiltoni.* The phase response curve so obtained is a curious one: it is bimodal, phase advances of up to 2 hours being produced by tidal stimulation before either of the two peaks in the activity pattern, and phase delays (of the same magnitude) by stimulation during or shortly after the activity peaks.

Enright (1961) found that *Synchelidium* would respond with an increased activity when the hydrostatic pressure was increased by less than 0.01 atmosphere. However, cyclical changes in pressure failed to entrain the activity rhythm and were eliminated as a natural Zeitgeber. In the green shore crab *Carcinus maenas,* on the other hand, changes in hydrostatic pressure as low as 0.1 atmosphere above that of the air are sufficient to entrain (Naylor and Atkinson, 1972). Cycles of 6.2 hours at atmospheric pressure followed by 6.2 hours at raised pressure, for example, were found to initiate tidal rhythmicity in previously arrhythmic crabs. When these crabs were subsequently transferred to constant conditions, the induced rhythm persisted for about 4 days. The greater the number of pressure cycles applied also increased the "precision" of the activity peaks so induced.

Semilunar rhythms

Biological rhythms with a period close to 14–15 days, and synchronized in nature to the semilunar or fortnightly cycle of spring or neap tides, are found in a number of littoral organisms, both plant and animal. These rhythms are sometimes called *circasyzygic,* the word *syzygy* referring to the astronomical conjunction when the gravitational pulls of the Sun and the Moon are combined (Chapter 1). The environmental synchronizers or Zeitgebers for such rhythms may be tidal or lunar.

One of the first semilunar periodicities to be reported was in the green flatworm *Convoluta roscoffensis* by Gamble and Keeble in 1903. These authors found that the colonies of this worm increased to a maximum during the spring tides, and then decreased to a minimum at the neaps; this periodicity was attributed to a cycle of reproduction. Another, possibly more famous, example is that of the Pacific grunion *Leuresthes tenuis,* a fish which rides high on the beach with the spring tides to spawn in the sand above mean high-water level. The eggs develop in the sand and hatch when agitated by the surf during the *next* spring high tide, 14–15 days later, whereupon the young fish are washed out to sea. In many cases the endogeneity of these semilunar rhythms is known. In the periwinkle *Littorina rudis,* for example, a species which inhabits the very highest part of the beach only reached by the spring high water, locomotor activity was found to persist in the laboratory with a roughly 15-day periodicity. Several other organisms show particularly intense activity at the times of the spring tides, despite the fact that they normally occupy the mid-tidal level. In two such examples, *Carcinus maenas* (Naylor, 1958) and the fiddler crab *Uca pugnax,* the circadian (~24-hour) and "circatidal" (~12.4-hour) components are thought to interact on the "beat" principle to produce the fortnightly semilunar increase in activity.

The brown alga *Dictyota dichotoma* shows a free-running and therefore endogenous semilunar rhythm of gamete discharge when maintained in the laboratory. The period of this rhythm is also somewhat *longer* than the spring tide to spring tide interval (14.7 days), and the phases of maximum spore discharge can be set in the laboratory by exposure to pulses of weak light simulating moonlight (Bünning and Müller, 1962). To test whether this periodicity was the result of an interaction between a circadian and a circatidal rhythm, Bünning and Müller entrained the circadian component to light cycles which differed from the normal solar day, with the expectation that concurrence between the two maxima should occur sooner with $T < 24$ hours, but later with $T > 24$ hours. The results showed that spore discharge occurred every 11-12 days when the algae were subjected to a light regime of LD *13.5:9.5* ($T = 23$ hours), and every 16-17 days when exposed to LD *14.25:10.25* ($T = 24.5$). These changes in the semilunar periodicity were in the expected direction, but of a smaller magnitude than expected if the tidal rhythmicity was unaffected by the LD cycle. Nevertheless, despite these discrepancies, this experiment offers some experimental support for the "beat" hypothesis. The green alga *Enteromorpha intestinalis* may also show a 14–15-day cycle of spore discharge, but the endogeneity of this system has not been established.

The most intensively investigated semilunar rhythmicity concerns the rhythm of emergence of *Clunio marinus,* one of the very few insects that can be described as truly marine. This minute (~2 mm) Chironomid midge has a most extraordinary life cycle. On the Atlantic coasts of France and Spain the larvae live at the *lowest* parts of the intertidal zone, and are only exposed at the times of spring low water. During certain of these exceptionally low tides the winged males emerge from their pupae, locate the

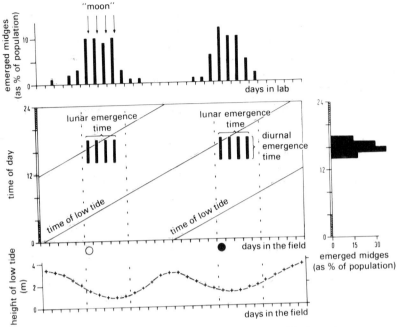

Figure 7.4 Semilunar rhythm of eclosion in the marine midge *Clunio marinus*.
Middle panel: emergence times in the field and the changes in low tide during one month.
Upper panel: emerged adults in LD *16*:8 with artificial moonlight (4 nights at 0.4 lux every 30 days).
Lower panel: Changes in the height of low tide during the month.
Right: Diurnal emergence time in LD *16*:8.
Note that emergence occurs at the time of spring low water, every 14–15 days. After Neumann, D. (1967), *Helgolander wiss. Meeresunters.*, **15**, 163–171, Fig. 2.

wingless females, assist them to emerge, and copulate. They then carry the females to the larval habitat where the eggs are laid. All this activity occurs during a two-hour period before the larval sites are covered by the incoming tide, and remain covered for the next 14 days.

On the Normandy coast the midges emerge only during the *evening* low spring water, never in the morning (figure 7.4). This behaviour suggests that tidal factors are not timing eclosion, or the midges would come out during both morning and evening low waters.

In a series of elegant experiments Neumann (1966) demonstrated that the rhythm of adult emergence in *C. marinus* was governed by a combination of a *circadian* rhythm controlling pupal eclosion and a *semilunar* rhythm determining the onset of pupation. In laboratory populations eclosion occurred towards the end of the daily photoperiod (i.e. about 12 hours after lights-on in *LD 16*:8), thereby corresponding with the observed time in the field. Eclosion was arrhythmic in continuous light (*LL*) but transfer of cultures from *LD* to *LL*, or exposure of an *LL*-raised culture to a single dark period, initiated a rhythm of eclosion which free-ran with an endogenous period (τ) of less than 24 hours.

The semilunar rhythm of pupation which is superimposed on this circadian cycle was shown to be entrained by natural or artificial "moonlight". Cultures of *C. marinus* from Normandy were raised in *LD 12*:12 or *LD 16*:8 and then exposed to pulses of weak light (0.4 lux) during the dark period of the cycle, for 4–6 days at intervals of 30 days. This treatment initiated and entrained the semilunar rhythmicity (figure 7.5) which was absent from the control populations. The endogeneity of this component was demonstrated by exposing a population of larvae to a *single* period of "moonlight", after which the system free-ran for more than three cycles (each of 14.7 days)before all of the individuals in the population had completed their development.

Although the *dates* of the spring tides are the same at all localities on a particular coast, the phase of the tidal cycle relative to local time may show differences even at the same longitude. These local differences were reflected in the times of eclosion, and cross-breeding experiments between

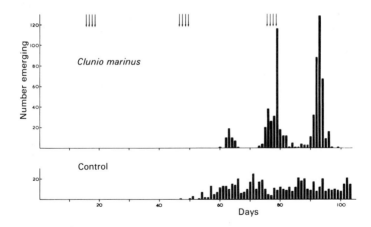

Figure 7.5 Semilunar rhythm of eclosion in *Clunio marinus* induced by artificial moonlight (4 nights with light at 0.4 lux every 30 days).
Above: experimental population;
below: controls without additional illumination at night. After Neumann, D. (1966), *Z. vergl. Physiol.*, **53**, 1–61, Fig. 12, Springer Verlag.

different local races established that eclosion time was inherited, involving a small number of genes.

In a more northerly locality (Heligoland, 54°N) the intensity of moonlight is apparently insufficient to act as a Zeitgeber, because the altitude of the moon is, on average, only 12.5° above the horizon, and the summer nights are considerably brighter than in areas further south. In this region, however, the 14–15 day periodicity was shown to be entrained by tidal stimulation, and probably controlled by an interaction between the circadian and tidal components. In the extreme north, above the Arctic circle (Tromsö) the larval habitat was found further up the beach and emergence was strictly tidal (every 12.4 hours), occurring with the initial exposure of the habitat during the ebb tide. Timing of this eclosion was found to be controlled by an "hour-glass" mechanism commencing at least 10 to 11 hours earlier during the *preceding* ebb.

Lunar rhythms

One of the best known of the lunar or *circasynodic* rhythms is undoubtedly that of the Palolo worm *Eunice viridis,* an inhabitant of coral reefs in the Pacific. This polychaete reproduces in October and November, and almost exclusively during a few nights in the last quarter of the synodical month. During these periods, the posterior sexual parts of the worms swarm on the surface of the water and liberate their genital products. The selective advantage of such behaviour is almost certainly the facilitation of reproduction. A related species *E. fucata* shows a similar type of behaviour in the Atlantic ocean, swarming on the surface during the third quarter of the lunar cycle in the months of June and July.

Swarming behaviour of the Mediterranean species *Platynereis dumerilii* has been studied in the laboratory, and its lunar rhythm shown to be endogenous (Hauenschild, 1960). In their natural habitats the worms become transformed into the sexual heteronereis form when sexually mature; they then rise to the surface, release their eggs and spermatozoa, then die. When maintained in the laboratory and in natural *LD* cycles, swarming was observed to occur in a peak close to the time of the new moon. When maintained in continuous light (*LL*), however, swarming was at random throughout the month. Rhythmicity was induced experimentally by interposing a few days at *LL* or *LD 18*:6 in every 30 days at *LD 12*:12, the peak of swarming then being phase-set by the end of the extended light period. Lunar rhythmicity could also be induced by providing additional nocturnal illumination in the form of very dim light (0.02 to 0.1

lux) for a six-day period every month. It is probable therefore that light of this intensity simulates natural moonlight and, as with *Clunio marinus,* moonlight is the Zeitgeber involved in circalunar entrainment.

Hauenschild noted no loss in swarming behaviour when moonlight failed to appear because of overcast night skies, and postulated that the lunar rhythm was endogenous. Laboratory experiments in which populations of worms were subjected to 30-day cycles of *LD—LL—LD,* and then maintained in an unchanging *LD*-regime, showed that the lunar periodicity could persist (for up to four months in one experiment), and with a free-running period (τ) of about 33 days. The persistence of the rhythm in unchanging photoperiod and the occurrence of a natural period *longer* than the synodical month, both indicate the endogenous nature of the rhythm in *P. dumerilii.* Worms maintained in a 23.5-hour light regime showed no shortening of the lunar period; this was taken as evidence against the hypothesis that the lunar period might arise from an interaction between a solar-day (~24 hour) rhythm and a second lunar-day (~24.8 hour) component.

Enright (1972) showed that the locomotor activity patterns of individual specimens of *Excirolana chiltoni* showed a lunar frequency in addition to the tidal frequency described earlier. This was expressed as a periodic "amplitude modulation" which paralleled in detail the complex lunar cycle of changes in tidal height. The free-running period of this bimodal circalunar component was, on average, one or two days *longer* than the natural 29-day lunar period, τ varying in different individuals from 26 to 33 days. The hypothesis that this circalunar rhythm arose as an interaction between a tidal (~24.8 hour) and a circadian (~24.0 hour) component was discounted.

Lunar rhythms are not confined to marine organisms, however, and several species of fresh-water and terrestrial insects are known to organize their lives on a monthly basis. The mayfly *Povilla adusta,* for example, emerges from the waters of Lake Victoria in its greatest numbers just after the full moon. Hartland-Rowe (1955) showed that this rhythm was maintained in the laboratory after the nymphs had been kept in the dark for 10 days, and in two individuals for six weeks. Working with the antlion *Myrmeleon obscurus,* Youthed and Moran (1969) showed that pit-building activity reached a maximum at the time of the full moon (figure 7.6). There was also a clear lunar-day (~24.8 hour) rhythm with a peak of activity about four hours after moonrise. The authors were able to demonstrate that the observed lunar rhythm (period about 28 days) was probably the result of an interaction between this lunar-day component

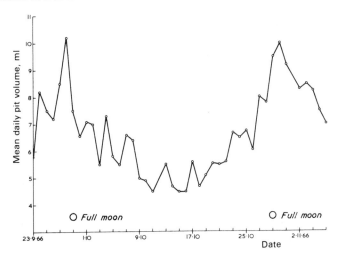

Figure 7.6 Mean daily pit volume of a group of 50 *Myrmeleon obscurus* larvae subjected to normal daylight conditions, and each larvae fed one ant a day. Open circles indicate times of the full moon. After Youthed, G.J. and Moran, V.C. (1969), *J. Insect Physiol.*, **15**, 1259-1271, Fig. 1, Pergamon Press, Oxford.

and a circadian rhythm. This was achieved in an experiment similar in principle to that employed by Bünning and Müller (1962) for the semilunar rhythm of spore discharge in the alga *Dictyota dichotoma*. Thus, twelve larvae of *M. obscurus* were exposed to a *reversed* light cycle in which the 14-hour dark period began at 9.45 a.m. instead of at its normal time in the evening. This treatment rapidly reversed the solar-day activity pattern, but the lunar-day activity peak still occurred about four hours after moonrise. Interaction between these two components now produced a lunar rhythm with a maximum at the time of the *new* moon, rather than at the full moon, a full half-cycle out of phase. The lunar rhythm of pit-building activity was also shown to free-run for at least two to three cycles (92 days) in DD, but to damp out in *LL*; it was therefore concluded to be endogenous. A lunar rhythm was absent, however, in larvae raised from the egg stage in the absence of moonlight.

CHAPTER EIGHT

BIOLOGICAL RHYTHMS: ANATOMICAL LOCATION OF PHOTORECEPTORS AND PACEMAKERS

CIRCADIAN RHYTHMS MAY BE OBSERVED IN SINGLE-CELLED ORGANISMS, in complex Metazoa, and at a population level. Without going into the concept of the circadian organization at the moment, suffice it to say that such "organization" probably consists of a "population" of autonomous or semiautonomous cellular clocks with a varying degree of mutual coupling, either via the environment (light and temperature cycles) or between the cells or oscillators themselves. Some of the evidence for this point of view has already been reviewed: this includes the dissociation of constituent circadian subsystems in a variety of organisms when kept in aperiodic environments or when treated with hormones (Chapter 2). Further evidence and its theoretical implications, particularly with respect to the circadian period, will be discussed in Chapter 9.

Despite the fact that circadian rhythmicity is probably a fundamental feature of cellular organization, circadian pacemakers at the tissue or organ level do occur. These are generally concerned with the control of specific behavioural or physiological functions involving a neural or hormonal output. It is hardly surprising, for example, that circadian pacemakers for locomotor activity or pupal eclosion, to name two examples, are to be found in the central nervous system or its associated structures. In other words, particular tissues and organs have evolved particular "clock" functions, just as tissues and organs have undergone differentiation to become livers, hearts or intestines. The discovery that specialized circadian "clocks" are localized within a particular part of the body, therefore, should not be confused with the concept of a "master clock".

Selected examples in which some progress has been made in locating the circadian pacemaker and its associated photoreceptor, are presented here; they include the "clocks" governing locomotor activity in cockroaches and birds, the eclosion "clock" in silkmoths, and the neural "clocks" of the sea hare *Aplysia*. The overt rhythmicity of a few of these clocks—notably those of the parieto-visceral ganglion and the eye of *Aplysia californica*—persist when isolated from the animal, thereby demonstrating some degree of autonomy. Most of the other evidence comes from experiments with intact animals, from extirpation and implantation, or from so-called "cut-and-see" procedures. In these experiments, the conclusion that a particular tissue or organ actually is or contains the photoreceptor and/or the clock must be subject to certain conditions or constraints. In the case of the photoreceptor, removal or occlusion of the involved organ should theoretically result in the *loss of entrainment* to an *LD* cycle, rather than arrhythmicity. Removal of the clock, on the other hand, should result in the *loss of the rhythm,* which in certain circumstances may be reinstated following the return of the organ to the body. It should always be kept in mind, however, that such experiments may only demonstrate than an essential "link" (between photoreceptor and clock, or between clock and its output, for example) has been removed, and that the correct controls and careful interpretation are essential.

Clocks controlling locomotor activity and other behavioural rhythms

In a series of papers between 1954 and 1960 Harker described experiments which claimed that the clock controlling locomotor activity in the cockroach *Periplaneta americana* was an endocrine pacemaker in the suboesophageal ganglion, and that the photorceptors involved in the environmental control of the rhythm were the dorsal ocelli or simple eyes (Harker, 1960). The most important of these experiments were:

1. *Parabiosis* between two insects, joined back to back with their haemocoels connected. In this experiment the top cockroach had its legs removed and had previously been kept in *LD 12*:12; it was therefore thought to be immobile and rhythmic. The lower cockroach was mobile and intact, but had been previously kept in *LL* and therefore thought to be arrhythmic (but see Chapter 2). The finding that the pair developed rhythmicity in *LL* was interpreted as showing that a "secretion, carried either in the blood or tissues, is involved in the production of a diurnal rhythm of activity in the cockroach".
2. In later experiments, using a headless (and therefore arrhythmic) cockroach as a recipient, it was claimed that implantation of the sub-oesophageal ganglion from a rhythmic donor caused the appearance of a "normal rhythm". The source of the supposed hormone was later traced to two pairs of neurosecretory cells on the ventrolateral surface of the ganglion. Harker claimed that these neuro-secretory cells could go on secreting with a circadian rhythm after all nervous connections had been broken; they were claimed, therefore, to be autonomous endocrine clocks.
3. Covering the compound eyes or cutting the optic nerves failed to "destroy the rhythm", but when the ocelli were painted over, the cockroach showed a gradual loss in rhythmicity even in an *LD* cycle. She therefore interpreted these results as showing the light stimulus is perceived through the ocelli.

These experiments and their far-reaching conclusions were amongst the first to claim the localization of a driving oscillation in any animal system. However, they were soon questioned by a number of other workers who failed to corroborate her findings and came to other conclusions. Without going into this aspect of the problem in great historical detail, it can be said that the majority of investigators now argue that the "clock" is probably in the optic lobes of the brain, that its output is neural (electrical) rather than hormonal, and that the compound eyes are the photoreceptors involved in entrainment, not the ocelli. The most important of these investigations will be outlined below; more complete reviews of the subject have been written by Brady (1969, 1974).

Harker's results and experimental designs were criticized by a number of workers including Roberts (1965), Brady, and Nishiitsutsuji-Uwo and Pittendrigh (1968). These criticisms included:

1. In parabiosis experiments it is essential that the recipient should be arrhythmic (i.e. not kept in *LL*, in which rhythmicity persists, Chapter 2).
2. Procedures should be adopted to prevent or take into account the possibility that the rhythm is transferred from the upper to the lower individual by mechanical means.
3. Explicit information on the respective *phases* of the two animals is needed: the two individuals should be purposely kept *out of phase* with each other before the experiment, and a transfer of phase, as well as a transfer of rhythmicity should be noted, before concluding that any transfer has occurred.
4. "Long" activity records are essential to overcome the trauma often observed during the first few days after any surgical procedure. In this connection it is interesting to note Brady's suggestion (1969) that running wheels tend to intensify the locomotor activity rhythm by some sort of positive feed-back, whereas in other recorders, such as rocking aktographs, operated animals may merely lie inactive after surgery.
5. All surgical procedures should be followed up with *post-mortem* histological examination to ascertain the extent and success of any section or ablation.

Roberts (1965) failed to reproduce any of Harker's important results, but found that a bisection of the brain caused arrhythmia, thus focusing attention on the brain and away from the sub-oesophageal ganglia. He also showed that painting over the compound eyes of *Leucophaea maderae* and *P. americana* with a mixture of lacquer, bees-wax and carbon black caused the insects to *free-run* in an *LD 12*:12 cycle (i.e. entrainment was lost). Subsequent removal of the paint resulted in entrainment to the light cycle, which was not destroyed by the surgical removal of the ocelli (figure 8.1). Although direct photo-stimulation of the brain was not excluded, these results indicated that

1. The brain was at least the site of an important link between the pacemaker and the thorax.
2. The compound eyes were the photoreceptors involved in entrainment.

Brady also failed to confirm Harker's major conclusions and showed

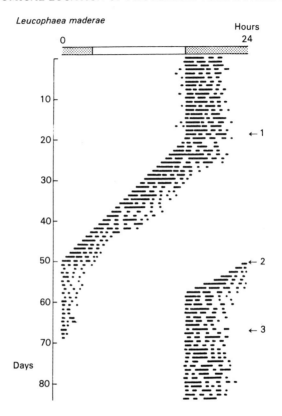

Figure 8.1 The role of the compound eyes in the entrainment of the locomotor activity rhythm of the cockroach *Leucophaea maderae*. The insect was kept in *LD 12*:12 throughout. At (1) the compound eyes were covered with an opaque black lacquer and entrainment to the light cycle was lost. At (2) the paint was pealed off and, after a series of advancing transients, the rhythm re-entrained. At (3) the ocelli were surgically removed with no effect. After Roberts, S.K. (1965), *Science*, **148**, 958–959, Fig.1, American Association for the Advancement of Science, Washington.

that cockroaches were able to maintain a normal rhythm after massive, if not total, reduction of their complement of median neurosecretory cells. Microcautery of the four cells in the sub-oesophageal ganglion, implicated by Harker as the site of the clock, also failed to destroy rhythmicity.

In a follow-up of Roberts' experiments, Nishiitsutsuji-Uwo and Pittendrigh (1968) found that bilateral section of the optic nerves caused *L. maderae* to free-run in *LD 12*:12, but that cutting the optic tracts (between the optic lobes and the rest of the brain) resulted in *arrhythmicity*. They

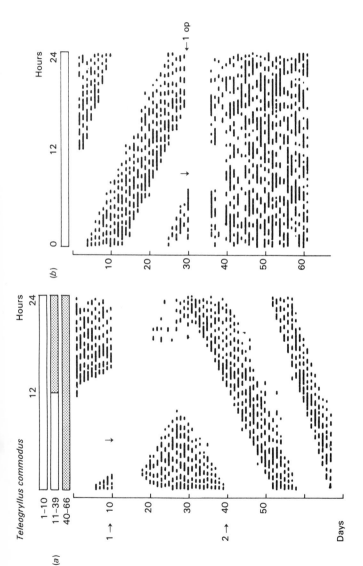

Figure 8.2 Stridulatory activity in the cricket *Teleogryllus commodus*.
(*a*) The insect was initially in *LL* and free-ran with τ >24 hours. At (1) the optic nerves were cut on both sides and the insect transferred to *LD 12:12*. After a period of silence, activity resumed but free-ran with τ > 24 hours as if the insect was in DD. A transfer to DD at (2) had no effect. (*b*) The insect was maintained in *LL* throughout. At (1) the optic tracts were cut on both sides. After a period of silence, stridulatory activity was resumed but was arrhythmic. After Loher, W. (1972), *J. comp. Physiol.*, **79**, 173–190, Figs. 6, 7, Springer Verlag.

concluded, therefore, that the "clock" was located in the optic lobes, and that these structures require connection to the compound eyes for entrainment to the light cycle, and to the rest of the brain, and hence to the central nervous system (CNS) and the legs, for the mediation of the locomotor rhythm. Essentially similar results have since been obtained for the stridulatory rhythm in the cricket *Teleogryllus commodus* (Loher, 1972); these results are illustrated in figure 8.2. Roberts (1974) later attempted to locate the driving oscillation to a more precise site within the optic lobes; he concluded that the two innermost elements (the lobula and the medulla) were crucial in the control of rhythmicity, whereas the outer synaptic area (the lamina) was essential in the coupling between the photoreceptor and the clock. It is interesting that if the circadian pacemaker is in the optic lobes, the whole system (photoreceptor-clock) is duplicated, and this raises important questions about their mutual coupling and integration of their output, perhaps in the protocerebrum.

Since practically all of the endocrine tissue can be cut away without

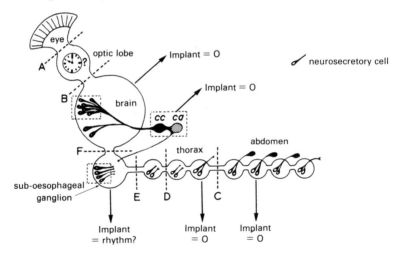

Figure 8.3 Synopsis of experiments on the control of the circadian rhythm of locomotor activity in cockroaches. The ganglia of the CNS are represented by the linked spheres, with neuroendocrine tissue, including known neurohaemal organs, indicated in black. Dotted boxes represent endocrine tissue which can be removed without altering the rhythm. Arrows show organs transplanted from rhythmic donors to headless arrhythmic recipients: 0 signifies that the host shows no detectable rhythm. Heavy broken lines are cuts made in the nerve trunks: cuts B,E,F, or splitting the protocerebral lobes bilaterally, apparently stop the rhythm; cuts A, D, C, or splitting the pars intercerebralis mid-sagittally, do not, *cc*—corpora cardiaca; *ca*—corpora allata. After Brady, J. (1971), in *Biochronometry,* ed. Menaker, M., pp. 517–526, Fig. 1, National Academy of Sciences, Washington.

upsetting rhythmicity, Brady (1969) and Roberts (1974) conclude that the locomotor activity clock in cockroaches has a neural (i.e. electrical) output. If that is so, then cutting the circum-oesophageal connectives (figure 8.3) should produce arrhythmicity. This result was obtained by Roberts at least for *L. maderae*.

It should not be supposed that the simple story outlined here is either complete or of general application, however. Evidence of a blood-borne factor (not necessarily a hormone) was obtained from careful parabiosis experiments with *P. americana* and the cricket *Acheta domestica* (Cymborowski and Brady, 1973). Species differences in the site and nature of the photoreceptor are also possible. For example, although Nishiitsutsuji-Uwo and Pittendrigh (1968) found that *L. maderae* with both optic nerves cut and with a glass 'window' inserted into the top of the head failed to entrain, *direct* photostimulation of the brain (or extra-optic photoreceptors) in some species appears possible, particularly where higher light intensities are used. A critical examination of some of these data is contained in a review by Brady (1974).

As in the cockroach, both endocrine and electrical mechanisms have been proposed for the control of locomotor rhythmicity in circadian and circatidal rhythms in crustacea. For over 30 years it has been known that eyestalk removal may cause a change in the *level* of activity. It is also known that the eyestalks contain important endocrine tissue (the X-organ), and the neurohaemal organs responsible for the release of their neurosecretory product into the haemolymph (the sinus gland). Since the injection of an eyestalk extract from donor crabs (*Carcinus maenas*) into eyestalkless recipients was found to cause more of a reduction in activity when the donors were in their low-tide (less-active) than in their high-tide (more-active) condition, a humoral mechanism for rhythm control has been proposed (Naylor and Williams, 1968). On the other hand, however, the eyestalks also carry the equivalent of the insect optic lobes, and studies with the crayfish (*Cambarus* sp.) have suggested that cutting the eyestalk nerve has the same effect as eyestalk removal, whereas injection of extracts or implantation of whole eyestalks into the haemolymph has none. Brady (1974) has therefore suggested that the situation is analogous to that in the insects, in that the pacemaker lies within the inner layers of the eyestalk ganglion and is coupled electrically (or at least axonally) to the effector organs (the legs). The role of the hormones is perhaps analogous to the effect of the corpora cardiaca in insects, affecting the *level* of activity, rather than its periodicity.

An attempt to locate the clock and the photoreceptor involved in the

Passer domesticus

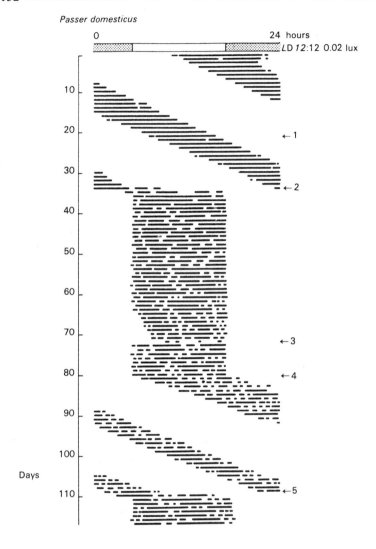

Figure 8.4 Locomotor activity rhythm in the house sparrow *Passer domesticus*. The bird was blinded and kept throughout in *LD 12*:12 with a light intensity of 0.02 lux. At (1) feathers were plucked from the bird's back. At (2) feathers were plucked from the bird's head, and the rhythm became entrained to the light cycle. At (3) the feathers, which had now regenerated, were again plucked from the head. At (4) India ink was injected under the skin of the head and the rhythm free-ran. At (5) some of the head skin, together with the ink, was scraped from the skull. After Menaker, M. (1971), in *Biochronometry*, ed. Menaker, M., pp. 315–332, Fig. 5, National Academy of Sciences, Washington.

rhythm of locomotor activity in birds has also been made (Menaker, 1971). In the house sparrow *Passer domesticus,* for example, the rhythm of perch hopping can be entrained—even in birds which have been blinded—to a light cycle in which the intensity of the (green) light is as low as 0.1 lux. This shows that the eyes are not the photoreceptors involved in entrainment. Some information *is* conveyed by the eyes, however, because sighted birds become arrhythmic in bright (50–500 lux) *LL* whereas blinded birds do not. The location of the extraretinal photoreceptor, somewhere within the brain, was demonstrated by the elegant experiment illustrated in figure 8.4. A blinded bird was exposed to an *LD 12*:12 light cycle with a sub-threshold light intensity (0.02 lux) to which the bird did not entrain. When a few feathers were plucked from the top of the head, however (a procedure which increases the amount of light reaching the brain by several orders of magnitude), entrainment to the *LD* cycle occurred. This entrainment subsequently failed as the feathers regenerated, and the bird again free-ran; replucking the feathers, however, again led to entrainment. Subsequent injection of India ink beneath the skin of the head (a procedure which reduces the amount of light reaching the brain by a factor of 10) again caused loss of entrainment and free-running behaviour. Finally, however, scraping off some of the head skin and the underlying ink allowed more light to the brain and *re-entrainment* to the *LD* cycle. This experiment strongly suggests that *direct* photostimulation of the brain is involved in entrainment, but does not implicate any particular part of that organ.

The search for the site of the circadian pacemaker in birds has produced more complex results. Working with *P. domesticus* and the white-crowned sparrow *Zonotrichia leucophrys,* Gaston and Menaker (1968) demonstrated that the pineal organ is essential for the *persistence* of the locomotory rhythm (and that of body temperature) in constant darkness. The pineal organ is therefore an essential link between the clock and the motor centres, if it is not the site of the pacemaker itself. Pinealectomy, however, did not interfere with the entrainment of the perch-hopping rhythm to the *LD* cycle, although certain changes in the phase-angle (i.e. activity commenced earlier relative to "dawn") were consistently recorded (Gaston, 1971). In birds, therefore, a curious situation seems apparent, in which the pacemaker and the mechanism responsible for its entrainment may be spatially distinct.

Clocks controlling insect eclosion

Despite their small size, which makes surgery (of the type described for cockroaches) difficult, the evidence available for fruitflies (*Drosophila* spp.) suggests that the clock controlling their rhythm of pupal eclosion is situated within the head, and that the organized photoreceptors are *not* involved in the coupling of this clock to the environmental light cycle. For example, the compound eyes and ocelli are only fully differentiated in the adult instar, but the rhythm may be initiated and phase-set by light pulses applied to larvae of all ages (Zimmerman and Ives, 1971). Entrainment can also occur in an eyeless and ocelliless mutant of *D. melanogaster*. In addition, light falling on the anterior end of the pupa generates a phase-shift as great as that falling on the whole organism, whereas that falling on the posterior half generates none. Since the anterior half contains the brain, and the brain is the one organ that "survives" metamorphosis with the least reorganization, it is reasonable to assume that the brain is the location of the eclosion clock.

More progress has been made with pupae of giant silkmoths whose large size facilitates the surgical manoeuvres necessary for the elucidation of such problems. Mixed-age populations of *Antheraea pernyi* and *Hyalophora cecropia* emerge as moths in specific "gates", the former late in the light component of the daily cycle, the latter in the forenoon. In *A. pernyi* the rhythm is circadian, free-running in DD with a period (τ) close to 22 hours. Working with the pupae of *H. cecropia,* Truman and Riddiford (1970) showed that operations such as section of the optic nerves, extirpation of the developing compound eyes, or removal of the sub-oesophageal ganglion, corpora cardiaca or corpora allata, were inconsequential, the moths emerging in their normal gate. Removal of the brain, however, resulted in the loss of gating control, and hence in random or arrhythmic eclosion. Subsequent reimplantation of the excised brains into the abdomens of the decerebrated pupae restored rhythmicity. In a more extensive experiment the brains were removed from 20 pupae, and then reimplanted, into the heads of ten individuals, and into the abdomens of the other ten. The pupae were then placed in holes drilled in an opaque board in such a way that the two ends of the pupae were exposed to lighting regimes (*LD 12*:12) differing only in phase. The results showed that the photoperiod experienced by the *brain*, whether it was now in the head or in the abdomen, determined eclosion time (figure 8.5). These elegant experiments showed that both the photoreceptor required for entrainment *and* the clock responsible for timing eclosion were contained within the brain tissues. Control of the eclosion rhythm in silkmoths is thus quite different from the control of locomotor activity in cockroaches.

In a later experiment, brains were interchanged between the two species,

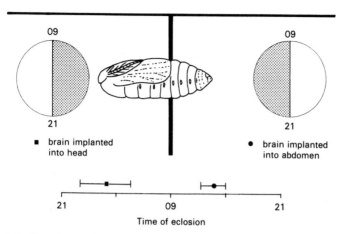

Figure 8.5 The eclosion of two groups of "loose-brain" *Hyalophora cecropia* which differed only in the site of brain implantation. The anterior end of each was exposed to light from 09:00 to 21:00; the posterior ends from 21:00 to 09:00. The time of eclosion was determined solely by the photoperiod to which the brain was exposed. After Truman, J.W. (1971), *Circadian Rhythmicity,* Wageningen, Fig. 3, Centre for Agricultural Publishing and Documentation, Wageningen, Netherlands (Pudoc). pp. 111–135.

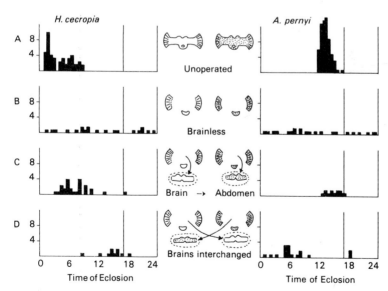

Figure 8.6 The eclosion of *Hyalophora cecropia* and *Antheraea pernyi* in an *LD 17*:7 regime showing the effects of brain removal, the transplantation of the brain to the abdomen, and the interchange of brains between the two species. After Truman, Fig. 2.

A. pernyi receiving the excised brain of *H. cecropia,* and *vice versa.* Results showed that the timing of eclosion was characteristic of the species of *brain,* but the emergence behaviour was characteristic for the *body* of the recipient (figure 8.6). Furthermore, since "loose-brain" animals functioned in the same way as intact animals, the results indicated that the output of the clock was hormonal; the hormone, moreover, was neither species- nor genus-specific.

Subsequent investigations showed that this hormone (the neurotropic ecdysis hormone) could be extracted from the brain of *A. pernyi* during the latter two-thirds of adult development within the pupa. It is secreted by the median neurosecretory cells of the brain and passed down their axons to the corpora cardiaca. The release of the hormone into the blood is a *gated* event dictated by the eclosion clock (Truman, 1971). Experimental injection of brain homogenates into moths competent to emerge, initiated a programme of abdominal movements which lead, within two hours, to eclosion, escape from the cocoon, and the spreading of the wings. The activation of this nervous programme by the hormone has been examined by electrophysiological methods. By recording from a nerve cord with severed peripheral nerves it was shown that the pre-eclosion behaviour was pre-patterned in the abdominal ganglia and required no sensory feedback.

It is interesting that pharate adults of *A. pernyi* exhibit essentially "pupal" behaviour until eclosion, restricting their movements to a simple rotation of the abdomen. This pupal behaviour persists even if the pupal cuticle is removed: "peeled" moths do not, for example, attempt to spread their wings. When the eclosion gate opens, however, the entire eclosion sequence is acted out, despite the fact that the insects have neither pupal integument nor cocoon to escape from. If the brain is removed, this sequence is destroyed. In some insects, for example, eclosion may occur before the resorption of the moulting fluid is complete and the moths emerge wet; in others the intersegmental abdominal muscles degenerate prematurely, so that the moth is trapped within its pupal cuticle. The proper sequence is restored, however, if a brain is implanted into the abdomen.

Photoperiodic clocks

Several early workers in insect photoperiodism found that blinding, or covering the compound eyes and ocelli with black paint, had no effect on the insects' ability to discriminate between long and short days. Experiments with light beams directed at different parts of the head, or with brain-transplant techniques, on the other hand, have indicated that *direct* photostimulation of the brain is involved. With respect to their location and to the nature of their photoreceptors, photoperiodic clocks in insects are therefore similar to those clocks controlling rhythms of eclosion or of other hormonally-mediated events. The same can probably be said for bird photoperiodism.

Lees (1964) maintained parent virginoparae of the aphid *Megoura viciae* at a short daylength (*LD 14*:10), but supplied a two-hour period of supplementary illumination to different areas of the head and body with fine microilluminators constructed from metal capillaries, or from plastic filaments (light guides) drawn from a viscous solution of polystyrene in benzene. The aphids were attached to the microilluminators for two hours each day; the rest of the time they were allowed to feed undisturbed on the host plant. The rationale behind this approach was that if the microilluminator was placed in a position to stimulate the relevant photoreceptor, the aphids would react to the long daylength (*LD 16*:8) so produced and give birth to further virginoparae, but if the supplementary light fell elsewhere, the aphids would respond to the unaltered short daylength and produce oviparous offspring. It was found that the most sensitive area was the dorsal midline of the head, directly above the brain. Light directed at the lateral parts of the head, or even directly into the compound eyes, was less effective; light directed at the thorax was completely ineffectual. Destruction of the compound eyes by microcautery had no effect upon the response. In this aphid species, ocelli are absent, but the cuticle overlying the brain is translucent. The brain, therefore, is probably fully accessible to light falling in this region.

In the Egyptian locust *Anacridium aegyptium,* a species with photoperiodic control of adult (ovarian) diapause, Geldiay showed that supplementary illumination directed at the midline of the head caused both oöcyte development and an increase in the diameter of certain neurosecretory cells of the brain. Short-day illumination of the whole body, or supplementary light exposure of the abdomen, on the other hand, had no such effect. It was suggested that long days activate the neurosecretory cells in the pars intercerebralis of the brain which, in turn, exert a positive control over the development of the ovaries.

The most persuasive evidence that both clock and photoreceptors are brain-centred has been obtained from surgical experiments performed with the giant silkmoth *Antheraea pernyi* (Williams and Adkisson, 1964). This moth passes the winter in a pupal diapause, but resumes its development in the spring when days exceed about 15 hours. Williams and Adkisson plugged diapausing pupae of *A. pernyi* into holes drilled in opaque board, and then exposed one side of the board to a diapause-maintaining photoperiod (*LD 8*:16) and the other side to a diapause-terminating photoperiod (*LD 16*:8). After seven weeks' exposure, all of the pupae with their head-ends exposed to long days had broken diapause, whereas all of those with their head-ends in short days had not. They then removed the brains from a further group of pupae and transplanted the excised brains into the tip of the abdomen, where they were covered with a transparent plastic window. This time they found that the photoperiod experienced *by the brain*, whether in the abdomen *or* in the head, was the decisive factor in controlling diapause termination. Subsequent experiments, in which small pieces of the brain were cut away before its implantation, showed that moths could distinguish between long and short daylengths, even when most of the brain tissue surrounding the pars intercerebralis had been removed. It was concluded that the mechanism was located in a "tiny mass situated just lateral to the medial neurosecretory cells".

The synthesis of brain hormone continues in diapausing pupae, the richest source of this hormone being those dormant pupae maintained in short days. The photoperiodic mechanism must therefore facilitate or control the translocation of brain hormone from the neurosecretory cells to the corpora cardiaca, and probably its release into the blood. Williams showed that pupae of *A. pernyi* could also distinguish between long and short days, even after injection of the puffer fish toxin, tetrodotoxin, which is thought to block action potentials and thereby "silence" the nervous system. Even at doses of 0.05 μg/g, sufficient to paralyse the developing insect and prevent its eclosion, discrimination between *LD 12*:12 (diapause-maintaining) and *LD 17*:7 (diapause-terminating) could occur, and development, in the latter, proceed unchecked to a fully developed pharate adult. The output of the photoperiodic clock is therefore clearly hormonal, and not dependent on the electrical properties of the axons.

One of the most intriguing aspects of photoperiodic control in silkmoths is how the light reaches the brain. The pupae (of *A. pernyi* and other silkmoths) are enclosed in a relatively opaque cocoon, but the pupal cuticle overlying the brain contains, in some species, a clear area of integument rather like a "window". Intuitively we would suspect that this facial window facilitates the transmission of light to the brain, and one Russian report states that painting over this area eliminated photoperiodic control. However, in repeated experiments with *A. pernyi,* Williams failed to substantiate this claim and concluded that if the window had such a role, it merely acts as a "safety device for cocoons in shady situations". The quality and the intensity of the light reaching the brain is considered elsewhere (Chapter 5).

Less-substantial progress has been made in avian photoperiodism, although direct photostimulation of the brain in the domestic duck was indicated in experiments performed by Benoit over 30 years ago. The existence of extra-retinal photoreceptors was confirmed by Menaker and Keatts (1968) who showed that bilaterally blinded (enucleated) house sparrows could discriminate between short daylength (*LD 6*:18) and long daylength (*LD 16*:8). They concluded that the photoreceptors, probably within the brain, are coupled via the clock to neuroendocrine centres controlling the annual reproductive cycle.

Later experiments demonstrated that the eyes in *Passer domesticus* play no part in photoperiodism (Menaker, 1971). Postrefractory and therefore photosensitive birds were left *with their eyes intact* and maintained for 39 days at long daylength (*LD 16*:8) in which the light intensity was close to the threshold for testis growth in untreated birds (10 lux). Two experimental groups were also exposed to the same light regime: in one, some of the feathers were plucked from the top of the head, in the other the skin of the head was injected with India ink. Thus the brains of the plucked group of birds were exposed to light intensities *above* the threshold, the brains of the injected birds were exposed to intensities *below* the threshold, and the eyes of both groups (and the controls) were exposed to light intensities *at* the threshold. If the eyes were involved in the photoperiodic response, both experimental groups and the control should have shown some testis growth. The results, however, showed that testis growth occurred only in the plucked birds, not in the controls, nor in those injected with ink. Light must therefore pass through the dorsal aspect of the cranium (as with the entrainment of locomotor activity) and impinge directly on the brain. The presence or absence of the eyes is inconsequential. The site of the extraretinal photoreceptor(s) within the tissues of the brain has not been determined.

Although there is no evidence that the "organized" photoreceptors are involved in insects or in birds, in the mammals retinally perceived light does exert at least partial control of the reproductive state. In rats, for example, constant light suppresses activity of the melatonin-forming enzyme (hydroxyindole-O-methyltransferase) in the pineal organ, whereas constant darkness increases its activity. Melatonin, in turn, inhibits gonadal function. Exposure to *LL* therefore stimulates reproductive activity by inhibiting the formation of melatonin. At least part of this response to environmental light is initiated in the retina, since blinding suppresses the effect of light on the enzyme (Moore *et al.*, 1967).

Autonomous neural clocks

None of the clocks discussed above show persistent circadian activity after removal from the body of the animal. This remarkable property, however, is shown by at least two organs from the sea hare *Aplysia californica*.

Strumwasser (1965) demonstrated that rhythms of electrical activity could be recorded from a single large (0.5 mm) neuron in the parieto-visceral ganglion, even when this ganglion was isolated from the animal and maintained in seawater or in a culture medium for several days. The particular neuron is a neurosecretory cell and was called the "parabolic burster" because of its characteristic bursting spike pattern. Strumwasser found that the output was basically *circadian*, with a peak of activity close to dawn, but a fortnightly (*semilunar*) and *annual* modulation of the output was also detected. The circadian output of the excised ganglion could be entrained by *LD* cycles, and in response to a shift in the driving light cycle the rhythm achieved steady state after a series of overshooting transients. This resetting by light pulses, even in the isolated ganglion, suggests direct photostimulation of the ganglion, and that the eyes are not necessary for entrainment, except perhaps at very low light intensities. A peripheral pacemaker can also be ruled out, as can a cyclical hormone or blood-borne agent emanating from another part of the body. The "parabolic burst" seems to be generated as a result of mechanisms endogenous to this single neuron, although interactions between adjacent neurons within the ganglion have not been rigorously excluded.

Lickey *et al.* (1971) later demonstrated that the electrical rhythm from this neuron could be entrained to *LD 13.5*:13.5 ($T = 27$) or to *LD 10.5*:10.5 ($T = 21$) as easily as to *LD 12*:12 ($T = 24$), at least for animals collected between November and March. Entrainment to non-24-hour Zeitgebers was quite unreliable at other times of the year. They also described a seasonal modulation in the steady-state phase angle adopted by the oscillator in a regime of *LD 12*:12. In this regime the oscillator was capable of assuming two stable phase relationships separated by 180°. Between April and October, for example, the two peaks occurred near subjective midday or midnight, but during the rest of the year (November to March) peaks were closer to dawn and dusk.

The second autonomous neural clock in *Aplysia* is the eye which continues to show a circadian rhythm of electrical activity from the optic nerve when removed from the animal and maintained in organ culture (Jacklet, 1969). The isolated eyes showed all of the recognized properties of a circadian system as outlined in Chapters 2 and 3. For example, the compound trains of action potentials persist in DD or in *LL* with an endogenous period differing only slightly from 24 hours, but sufficiently so that the free-running rhythm rapidly becomes out of phase with environmental time. The system also obeys "Aschoff's rule" (for a diurnal animal) in that τ in DD is longer than τ in *LL*. The rhythm is also entrained by photoperiod, either in the intact animal, or *in vitro,* and phase-shift experiments with light pulses can produce either advances or delays, according to the phase of the oscillation perturbed, frequently after a series of non-steady states or transients. These observations demonstrate that the photoreceptor, the oscillator, and the mechanism for phase adjustment all reside within the eye, and thus form an autonomous circadian system. It is also of interest that although the eyes are *not* required for the entrainment of the parabolic burster, they are necessary for the entrainment of the rhythm of locomotion.

CHAPTER NINE

THE FUNDAMENTAL NATURE OF BIOLOGICAL RHYTHMS

CIRCADIAN RHYTHMICITY IS GENERALLY CONSIDERED TO BE A FUNDA-mental aspect of cellular physiology, and tacitly assumed to be a feature of eukaryotes, perhaps with a mechanism common to them all. Neverthe-less, despite the not inconsiderable amount of work applied to "fundamen-tal" aspects of circadian rhythmicity, the available evidence provides only hints or guidelines for future research, rather than a concise and unequivo-cal account of the concrete mechanisms involved. Some work, for example, suggests that the nucleus and protein synthesis are closely connected with (but not an integral part of ?) the circadian oscillator, whilst other work indicates that membrane physiology, cytoplasmic organelles, or higher-frequency biochemical oscillations are involved. Since circadian rhythms have not been recorded from isolated sub-cellular particles or from cell-free extracts, it is probable that the "biological clock" does not lie *within* the cell, but *is* the cell, the spatial integrity of the cell and interactions between organelles being essential for the special features of circadian rhythmicity.

Much of the experimental work has been carried out with unicellular organisms, particularly algae, with the justifiable hope that these "simple" systems offer the most favourable material for the solution of fundamental problems. The techniques adopted have mainly concerned the chemical manipulation of their circadian rhythms in free-running conditions. In this connection it is of paramount importance to distinguish between treat-ments which affect the "clock" itself and those which merely affect the "hands of the clock". For example, chemical perturbations which alter the *period* of the free-running oscillator, or bring about stable *phase-shifts,* are clearly affecting the clock and/or its associated phase-resetting mechan-ism, whereas treatments which merely affect the *amplitude* of the rhythm, or even abolish it, may only be affecting the overt process or driven rhythm. To quote examples of the latter, Hastings (1970) found that the photosyn-thetic inhibitor DCMU (dichlorophenyl dimethyl urea) blocked the rhythm of photosynthesis in the alga *Gonyaulax polyedra,* but after its removal the rhythm continued *in phase* with the rhythm before treatment. Similarly, the antibiotic puromycin inhibited the rhythm of spontaneous glow but, once again, the rhythm reappeared in phase after the drug's removal. In this organism it is clear that photosynthesis and luminescence are not an integral part of the circadian pacemaker, but merely overt events driven by the clock. Even in those organisms in which cell division is "gated", the cell cycle is distinct from the clock because chloramphenicol halts the cell cycle whilst other rhythmic events continue.

This chapter will also discuss certain other fundamental aspects of biological rhythms, including temperature-compensation, the circadian period, and the nature of "circadian organization".

The role of the nucleus and protein synthesis

The marine alga *Acetabularia* has become a favourite subject for the study of nuclear-cytoplasmic "relationships", including that possibly involved in circadian rhythmicity, because of the relative ease with which the nucleus can be removed and transplanted. This alga is both unicellular and uninuclear, but grows to a length of about 5 cm. Each cell consists of a stalk, a rhizoid which contains the nucleus and, at maturity, an apical cap which differs in morphology between the various species. Several circadian rhythms have been described. These include oxygen evolution and photosynthesis, events which can be monitored, even in single cells, by means of platinum electrodes, Cartesian diver techniques, or by classical Warburg manometry.

Early observations showed that the nucleus was not essential for the maintenance of rhythmicity. Working with *A. major,* for example, Sweeney and Haxo showed that the rhythm of oxygen evolution could be observed in both nucleate and enucleate cells. This was confirmed by Schweiger (1971) who showed that enucleate cells of *A. mediterranea* could retain their circadian rhythmicity for over six weeks in constant conditions (weak *LL*). It was shown, however, that the nucleus does influence *phase*. Two cells were entrained to *LD* regimes in 180° antiphase and then released into *LL*. The rhizoids containing the nuclei were then interchanged between the two cells, and the phases of the photosynthetic rhythm in each observed to shift by a full half-cycle. An essentially similar result was achieved with nuclear transplantation, showing that the nucleus alone was responsible for the phase-shift, rather than some other constituent of the rhizome transplanted along with it (figure 9.1a). In another experiment, the basal and apical parts of the cells were exposed for 14 days to light regimes

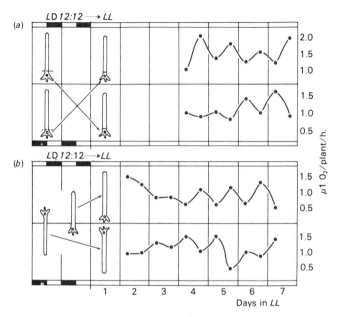

Figure 9.1(a)   Showing that transplantation of the nucleus-containing rhizoid from one cell of *Acetabularia* to another results in a 180° shift in phase when the two cells are initially entrained to light cycles in antiphase. The time of highest oxygen output is dictated by the phase of the nucleus.
(b) Showing that illumination of the nuclear area in *Acetabularia* determines phase. After Schweiger, H.G. (1971), *Circadian Rhythmicity,* Wageningen, pp. 157–174, Figs. 5, 7.

180° out of phase, before release into *LL*. This experiment showed that the cytoplasmic rhythm (of photosynthesis) was phase-set by illumination of the nuclear area (figure 9.1b). The partial "independence" of the cytoplasm in *Acetabularia,* however, does not mean that DNA-dependent RNA and protein synthesis is not involved, since the chloroplasts possess their own DNA and are able to synthesize ribosomal and transfer RNA in the absence of the nucleus.

The second experimental approach has been the use of various antibiotics which selectively inhibit various stages of RNA and protein synthesis. However, this seemingly promising field has produced somewhat perplexing and sometimes conflicting results with few clear indications of the exact role that protein synthesis might have in the circadian system.

In the marine dinoflagellate *Gonyaulax polyedra* there are endogenous circadian rhythms of photosynthesis, cell division, luminescent "glow" and of stimulated flashing, all of which can be entrained by an *LD* cycle, but with each attaining its characteristic phase-angle. The rhythms disappear in bright *LL*, but persist in weak *LL* and, for a time, in DD. The free-running periods range from 22 to 27 hours with a small, often negative, Q_{10} (see Chapter 2). Hastings and his associates (see Hastings and Keynan, 1965) have reported the effects of a whole range of inhibitors on these systems. Chloramphenicol, for example, which is thought to inhibit protein synthesis on 70S ribosomes, had no effect on phase or period, although it caused an increase in amplitude. Puromycin, which is a potent inhibitor of protein synthesis on both 70S and 80S ribosomes, also affected only the amplitude of the rhythm, which it depressed. These results indicated that protein synthesis at the *translation* level plays no essential part in circadian rhythmicity. The most interesting effects were noted with actinomycin-D, an inhibitor of the *transcription* of nuclear and extra-nuclear DNA by specifically binding to the guanine residues of the DNA

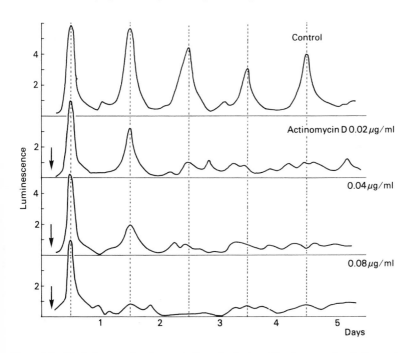

Figure 9.2 The influence of actinomycin D on the rhythm of luminescent glow in the marine dinoflagellate *Gonyaulax polyedra*. Arrows show time at which the inhibitor was added. Note that actinomycin at 0.08 μg ml[-1] allows the first peak of luminescence to occur before the rhythm is extinguished; lower concentrations allow two peaks. After Karakashian, M. and Hastings, J.W. (1962), *Proc. natn. Acad. Sci.*, **48**, 2130.

molecule. In cells of *G. polyedra* (figure 9.2), and in *Acetabularia*, it had no effect on period or phase, but reduced the amplitude of the rhythm and finally obliterated it. In *G. polyedra* it stopped the rhythm of luminescent glow but not that of photosynthesis or of stimulated flashing; for these reasons it has been concluded that actinomycin has no effect on the circadian pacemaker, and that transcription as well as translation plays no essential part in the mechanism. However, following treatment with the inhibitor, the first peak of luminescence was not affected, indicating that determination occurs at least 24 hours before its expression, and that cells seem to synthesize a "species" of RNA which "commits" their programme. At low concentrations of actinomycin (0.02 μg ml^{-1}) two or more such peaks may occur before the rhythm is extinguished.

The one clear demonstration of an inhibitor modifying the clock itself comes from studies on *Euglena gracilis* treated with cycloheximide (Feldman, 1967). In these experiments the cells were grown in LD *12*:12 until the late logarithmic phase and then transferred to DD in which the persistent rhythm of phototaxis was measured by the cells' response to a test light turned on for 24 minutes every two hours. Feldman found that the addition of cycloheximide (actidione), an inhibitor of protein synthesis at the cytoplasmic ribosomes, particularly in eukaryotic cells, caused a clear concentration-dependent lengthening of the free-running period. In the control cultures, for example, τ was about 23.7 hours, but τ became 25.0 hours after the addition of cycloheximide at a concentration of 0.2 μg ml^{-1}, 28.0 hours at 1.0 μg ml^{-1}, and about 36.0 hours at 4.0 μg ml^{-1}. These increases were stable for at least five cycles, but were reversible (to about 24 hours) when the antibiotic was washed out of the cultures.

Vanden Driessche (1971) found an endogenous rhythm in the incorporation of labelled uridine into RNA fractions from enucleated *Acetabularia*, and thereby directed attention towards the role of the chloroplasts. Apart from RNA synthesis, several other circadian rhythms associated with the chloroplasts are known. These include ATP content, chloroplast shape, polysaccharides, number of carbohydrate granules, and division. All of these can be observed in enucleate cells and are known to be endogenous, with the exception of the rhythm in the number of granules, which has only been studied in LD (Vanden Driessche, 1971). Although the chloroplasts in *Acetabularia* appear to possess an independent ability to synthesize RNA in a circadian fashion, some evidence for nuclear messenger controlling chloroplast rhythmicity was obtained by the use of RNAase: in enucleated cells, treatment with this enzyme caused a loss in the photosynthetic rhythm which was not observed in whole cells similarly treated. Nuclear

RNA may therefore play a central role, but the persistence of rhythmicity in enucleate cells demonstrates that *daily* transcription from nuclear DNA is not necessary, nor is the daily transcription from chloroplast DNA. In addition, daily translation cannot be involved in *Acetabularia* since chloramphenicol and puromycin only reduce amplitude, and have no effect on period or phase.

Despite these complications, and the conclusion that neither transcription nor translation are involved, theoretical models for circadian rhythmicity based on such events have been proposed. One which deserves attention is the *chronon* theory (Ehret and Trucco, 1967) which suggests that the genome of eukaryotic cells contains polycistronic elements (= chronons) needing about 24 hours to complete their transcription.

In summary, it seems that the various stages of protein synthesis are *not* the clockwork involved in circadian rhythmicity (with the interesting possible exception of *Euglena*), but that protein synthesis is involved in its expression. Indeed, it would be hard to imagine how the synthesis of cellular proteins, which include both structural elements and enzymes, is not somehow involved in such a fundamental aspect of cellular physiology.

Agents which alter period and phase

The two most important "clock" properties, period and shifts in phase, are remarkably resistant to chemical manipulation. This was demonstrated in the foregoing section in which a wide range of inhibitors of macromolecular biosynthesis were shown to be almost without effect on these parameters, the only exception being the period-lengthening effect of cycloheximide in *Euglena gracilis*. This apparent intractability of period, in particular, was the almost universal experience of earlier investigators. Bühnemann (1955), for example, working with the alga *Oedogonium cardiacum,* found that a wide array of chemicals including NaCN, 2–4 dinitrophenol, Na_2HAsO_4, NaF, quinine, $CuSO_4$ and cocaine had no effect on τ. Indeed, this remarkable property has become one of the major problems in circadian rhythmicity, along with the perhaps related question of temperature-compensation.

A number of chemical treatments are now known to lengthen τ, however, amongst which are heavy water (D_2O), ethanol, lithium salts, and the antibiotic valinomycin. Since more is known about the effects of D_2O we will deal with this agent first.

Working with *Euglena gracilis,* Bruce and Pittendrigh (1960) reported that τ for the rhythm of phototaxis in cultures transferred from H_2O to

$H_2O:D_2O$ mixtures, sometimes lengthened by several hours (τ in H_2O was 23.5 to 23.75; τ in D_2O was 26.5 to 28.0 hours). They also showed that this effect was reversible, τ returning to about 24 hours in cultures washed free from heavy water and returned to H_2O. Similar effects have now been reported, with circadian rhythms from higher plants such as the bean *Phaseolus,* the rodent *Peromyscus leucopus,* insects such as the fruitfly *Drosophila pseudoobscura* and the cockroach *Leucophaea maderae,* from birds, and the tidally-rhythmic isopod *Excirolana chiltoni.* Heavy water is also known to lengthen τ as revealed in "resonance experiments" with the photoperiodic oscillator in the flowering plant *Chenopodium rubrum* (Brenner and Engelmann, 1973). The fact that such similar effects are found in such a wide range of dissimilar organisms is concluded by most workers to be significant, perhaps indicating that this period-lengthening effect of heavy water will tell us something about the fundamental nature of the clock. This being likely, it is worth while examining some of these data in more detail.

The most complete data concern *Peromyscus, Excirolana* and *Droso-*

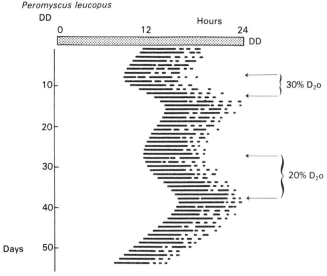

Figure 9.3   The effect of heavy water (D_2O) on the free-running rhythm of locomotor activity in the deer mouse *Peromyscus leucopus.* When supplied with H_2O in the drinking water, τ is less than 24 hours. Addition of 30 per cent and 20 per cent D_2O, however, lengthens τ. This lengthening effect is dose-dependent and reversible. After Suter, R.B. and Rawson, K.S. (1968), *Science,* **160,** 1011–1014, Fig.1, American Association for the Advancement of Science, Washington.

phila. Suter and Rawson (1968), working with the deer mouse *P. leucopus* and its running activity rhythm, found a clear concentration-dependent lengthening effect of D_2O on τ which appeared to have no threshold and was reversible, returning to about 24 hours soon after the D_2O was withdrawn from the drinking water. The maximum lengthening of τ was about 6 per cent when the drinking water contained 30 per cent D_2O (figure 9.3). Enright (1971a) found a remarkably similar effect on the tidal rhythm in *E. chiltoni*; again it was dose-dependent and reversible. Enright also compared the effects of heavy water on the circadian oscillation in *E. chiltoni* with a number of high-frequency (non-circadian) oscillators such as the electric-organ discharge in a Gymnotid fish, respiratory cycles in a goldfish, and cardiac cycles in a clam and a crab. He found that the frequency of all was decreased (i.e. slowed down), although to a *lesser* extent in the circadian rhythm. This difference was attributed by Pittendrigh *et al.* (1973) to the fact that the periods of circadian oscillators are temperature-compensated, whilst those of non-circadian rhythms are not. They then proceeded to show a similar differential effect of heavy water in *D. pseudoobscura* using a temperature-compensated parameter of the eclosion rhythm (τ) and a temperature-dependent one (the phase relationship between the median of eclosion and light-on). They concluded that the important clock properties, particularly τ, were not only temperature-compensated but "deuterium-compensated", and indeed compensated for all changes in the cellular *milieu* which might destroy their ability to measure time.

Much less is known about the other agents which affect period. Ethanol offered as a dilute solution in the transpiration stream lengthened τ by several hours in the bean seedling *Phaseolus*. Enright (1971b) demonstrated a similar effect, apparently dose-dependent up to 1 per cent ethanol, in lengthening the tidal rhythm of *Excirolana chiltoni*; at a concentration of 0.5 per cent τ was increased by about an hour. Engelmann (1972) found that the free-running period of the rhythm of petal movement in *Kalanchoë blossfeldiana* increased from 22.4 hours to 24.3 hours after the administration of lithium ions. Lastly it has been reported that the antibiotic valinomycin can induce phase-shifts in *Phaseolus* and *Gonyaulax* which closely mimic those produced by light.

The possible role of membranes

Heavy water, lithium ions and ethanol are all known or thought to influence ionic balances across cellular membranes, the former possibly because the increased mass of the deuteron slows *diffusion,* and lithium and alcohol by affecting membrane *permeability.* The antibiotic valinomycin is a highly specific carrier of potassium (K^+) ions, and these ions are known to cause stable phase-shifts in the isolated eye of *Aplysia,* similar to those of light. Many authors have therefore interpreted the period-lengthening effects of these agents as highly suggestive evidence that *ions and membrane permeabilities* are somehow involved in circadian rhythmicity, perhaps as part of the cellular clockwork itself. This point of view is strengthened by the fact that many higher-frequency biological rhythms (such as those examined by Enright, 1971*a*) are essentially membrane processes, and that several aspects of membrane physiology, and diffusion itself, are already temperature-compensated. Such a possibility would be consistent with the findings that inhibitors of protein synthesis were largely without effect on period or phase, although having an effect on the "expression" of the rhythms because proteins are involved in the structural elements (transport proteins) concerned. It might also be consistent with the observed role of cytoplasmic organelles such as chloroplasts and mitochondria. In higher animals with a complex nervous system (and identifiable circadian pacemakers in the CNS which control overt behavioural rhythms), ionic permeability could be involved in rhythmic (circadian) nerve activity.

A model for circadian rhythmicity based on a feedback mechanism between ion concentrations and the activities of membrane-bound ion transport elements has been proposed by Njus *et al.* (1974). In this model light is presumed to act by perturbing ion concentrations, probably K^+, which pass through photosensitive ion gates to augment or deplete the K^+ concentration directly. In higher systems, hormones could mediate between specialized photoreceptors and the membrane. Since the clock is envisaged as a feedback between ions and membranes, there must be circadian changes in transmembrane ion fluxes, probably involving the activation and inhibition of transport proteins in the membrane. One of the attractions of this model is that it can account for temperature-compensation of the circadian period. It is known, for example, that in biological membranes the transport proteins are intercalated in the lipid bilayer and move laterally in the fluid lipid matrix, and that the lipid composition of membranes is controlled to maintain a definite fluidity, independent of temperature. It is possible that "clock mutants" (Chapter 2) involve altered proteins which might be involved in ion transport or lipid metabolism, or in changes in this lipid fluidity.

Brinkmann (1971) has drawn attention to the possible role of the mitochondria. He noted that bacteria lack mitochondria *and* circadian rhythms, whereas fungi possess different types of mitochondria and diverse rhythms, only some of which are circadian. All other plants and animals, on the other hand, have a similar type of mitochondrion, and possess a characteristic temperature-compensated circadian rhythm. The implication that mitochondria (as well as chloroplasts) are involved, is probably consistent with the membrane-ion transport hypothesis.

The circadian period and circadian organization

One of the difficulties inherent in seeking an explanation for circadian rhythmicity concerns the *period* of the oscillator which, at about 24 hours, is several orders of magnitude greater than the known biochemical oscillations within the cell. The period is also temperature-compensated and extremely "accurate" in that its day-to-day variation is very slight. The problem with biological rhythms of even longer period (semi-lunar, lunar, annual, etc.) is even more difficult to explain. In the *chronon* model for circadian oscillators, the 24-hour period is attributed to the time taken for

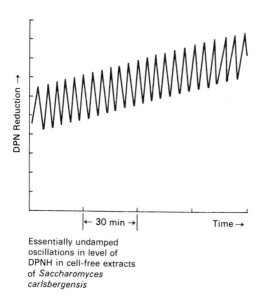

Essentially undamped
oscillations in level of
DPNH in cell-free extracts
of *Saccharomyces
carlsbergensis*

Figure 9.4   Undamped high-frequency biochemical oscillation (of level of DPNH) in a cell-free extract of the yeast *Saccharomyces carlsbergensis*. After Pye, K. and Chance, B. (1966), *Proc. natn. Acad. Sci.,* **55**, 888–894, Fig. 4.

the transcription of a hypothetical linear sequence of DNA, cistron by cistron. In the *membrane* model, it is ascribed to the slower processes involved in the lateral diffusion of proteins within the lipid bilayer. In a third model, briefly considered here, the circadian period is conceived to be the result of interactions between *higher-frequency biochemical oscillations* within the cell.

Such oscillations (figure 9.4) are well known in cells or in cell-free extracts (Pye and Chance, 1966). As noted above, however, most of them oscillate with a much higher frequency, usual periods being only of minutes or seconds. They are also strongly temperature-dependent, having Q_{10}'s between 2 and 4.

In a theoretical study Pavlidis (1969) has suggested that the circadian period might arise within a population of cellular biochemical oscillators as a result of strong coupling between them, even though in isolation each would oscillate at a much higher frequency. He constructed a mathematical model consisting of a number of hypothetical high-frequency oscillations which, under the influence of strong mutual coupling, showed a net frequency reduction of several orders of magnitude, from minutes to hours. The frequency of the population was found to be a function of the *number* of individual oscillators. It was later demonstrated, again in a mathematical model, that such a coupled system could show many of the features of circadian rhythmicity, such as temperature-compensation, a phase response curve, and "splitting" into separate subsystems. However, although the study suggests that circadian oscillations might have originated in high-frequency oscillations within the cell, we have little experimental evidence for such an interaction.

One experimental study has shown at least a remarkable similarity to the model proposed by Pavlidis: this concerns the isolated eye of *Aplysia californica* and its rhythm of neural activity (Jacklet and Geronimo, 1971). The retina of the *Aplysia* eye contains about 3700 receptor cells and about 950 secondary cells, the latter firing in synchrony with optic nerve activity. Jacklet and Geronimo found that surgical removal of these secondary cells caused little change in the free-running period of compound-action potentials, until about 80 per cent of the "population" had been removed. Thus, in the intact eye, the circadian period (τ) was 26–28 hours. When 80 per cent of the cells had been cut away, τ became 25 hours, with a minor mode at 20–22 hours. After a 90 per cent reduction, τ became 21 hours, with minor modes at 10 and 7.5 hours. Finally, when the number of cells was reduced to about 73 (a 98 per cent reduction) the major mode was at 7.5 hours with minor modes at 12–16 and at 18–20 hours (figure 9.5).

Jacklet and Geronimo considered three possible explanations for their data:

1. that the "population" of secondary cells were driven by a "master clock" showing a circadian oscillation,
2. that the "population" consisted of a population of interacting circadian oscillators,
3. that there was a population of non-circadian oscillators which together produce the circadian period as in Pavlidis' model.

The authors, however, pointed out that under hypothesis (1) the period should have stayed constant (at about 24 hours) until the master was cut away, and if (2) was the explanation, the minimal population should still have shown a circadian periodicity. They concluded, therefore, that hypothesis (3) was the most likely, and that the circadian period of the whole organ was a function of the *number* of constituent subsystems, each with a higher frequency. It should be noted, however, that this case involves *cells* and not sub-cellular biochemical oscillators.

No circadian rhythms have been described in sub-cellular fractions (cell-free enzyme systems oscillate with a much higher frequency, and are temperature-dependent), although they are commonplace, if not ubiqui-

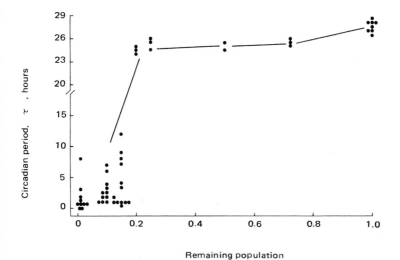

Remaining population

Figure 9.5   The shortening of the circadian period in the rhythm of neural activity in the isolated eye of the marine mollusc *Aplysia californica* as the "population" of "secondary cells" in the retina is surgically reduced. Eighty per cent of these cells may be removed before the period (τ) assumes a much higher frequency. When 98 per cent was cut away the period became 7.5 hours. After Jacklet, J. and Geronimo, J. (1971), *Science,* **174,** 299–302, Fig. 2, American Association for the Advancement of Science, Washington.

tous, in the intact cells of eukaryotes, both unicellular and multicellular. It is likely therefore that the eukaryotic cell does not *contain* the clock, but *is* the clock. This further suggests that multicellular animals and plants are "populations" of cellular clocks, and raises the question of the nature of the "circadian organization". Experimental evidence for this organization has been noted earlier in this book; perhaps the most persuasive of this is the "splitting" of activity rhythms into separate components in rodents, birds, and primates (Chapter 2).

In a theoretical study, Winfree (1967) demonstrated how a population of oscillators, all with a similar periodicity and a *weak* mutual coupling, tend to synchronize each other to a common phase and period, even though starting at random phase to each other. In this way the individual cellular circadian oscillators could become mutually entrained to provide the whole-organism synchrony observed in multicellular creatures. Using an electronic analogue, Winfree has also shown how a population of weakly interacting oscillators can undergo "splitting" as observed in animal experiments.

Since organisms or cultures of protists may show persistent circadian rhythms with little or no damping, it is reasonable to assume that some sort of intercellular communication occurs, such as a weak chemical coupling. (If it does not, then the circadian organization is more akin to a "clock-shop", Winfree, 1975.) Such intercellular communication has been sought in populations of *Gonyaulax polyedra* and in *Euglena gracilis* by mixing together populations of cells several hours out of phase, but none was found. In the endocuticle of many insects, however, "day-night" layers in chitin orientation are formed by the epidermal cells acting apparently in unison (Neville, 1970). Furthermore, in certain beetles, endocuticular layer formation—although in this case *non*-circadian—occurs at different frequencies in different parts of the body, although at the *same* frequency in the adjacent cells of one particular area. This suggests that some kind of cellular communication does occur in multicellular animals, similar perhaps to that which must occur in the eye of *Aplysia.*

Apart from the whole organism synchrony arising from coupled cellular oscillators, there are highly organized "circadian pacemakers" located in the central nervous system with highly specific functions controlling overt behavioural or developmental events (Chapter 8). In some cases these clocks are duplicated as a result of bilateral symmetry. At another level there is evidence of hierarchical entrainment: "driving" or A-oscillators coupled directly to the environmental light cycle, and "driven" rhythms or B-oscillators deriving their entrainment only indirectly through a driver

(Chapter3). In some insects there is evidence for more than one driver, each with distinct properties and with "responsibility" for a discrete function or set of functions. In still other organisms, biological rhythms with different periodicity may co-exist and perhaps interact (e.g. tidal with daily rhythms, daily with semi-lunar, daily with annual). The "temporal organization" of higher animals (and perhaps plants) becomes therefore, in this view, much more complex than at first suspected.

FURTHER READING

Chapter 1.

Aschoff, J., Saint Paul, U. von, and Wever, R. (1971) "The Longevity of Flies under the influence of Time Displacement" (in German). *Naturwiss., 58,* 574.

Brown, F.A. jr., Hastings, J.W. and Palmer, J.D. (1970), *The Biological Clock: Two Views,* Academic Press, New York and London.

Bruce, V.G. (1972) "Mutants of the Biological Clock in *Chlamydomonas reinhardi,*" *Genetics, 70,* 537–548.

Bünning, E. (1971) "The Adaptive Value of Circadian Leaf Movements," in *Biochronometry,* ed. Menaker, M., pp. 203–211, National Academy of Sciences, Washington.

Bünning, E. (1973) *The Physiological Clock,* Third English Edition, Springer Verlag.

Hamner, K.C., Flinn, J.C., Sirohi, G.S., Hoshizaki, T. and Carpenter, B.H. (1962) "Studies on the Biological Clock at the South Pole," *Nature, Lond., 195,* 476–480.

Hillman, W.S. (1956) "Injury of Tomato Plants by Continuous Light and Unfavorable Photoperiodic Cycles," *Am. J. Bot., 43,* 89–96.

Jacklet, J.W. (1969) "Circadian Rhythm of Optic Nerve Impulses recorded in Darkness from Isolated Eye of *Aplysia,*" *Science, Wash., 164,* 562–563.

Jerebzoff, S. (1965) "Manipulation of Some Oscillating Systems in Fungi by Chemicals," in *Circadian Clocks,* ed. Aschoff, J., pp. 183–189, North-Holland, Amsterdam.

Konopka, R. and Benzer, S. (1971) "Clock Mutants of *Drosophila melanogaster,*" *Proc. natn. Acad. Sci., U.S.A., 68,* 2112–2116.

Oatley, K. and Goodwin, B.C. (1971), in *Biological Rhythms and Human Performance,* ed. Colquhoun, W.P., Academic Press, London and New York.

Pavlidis, T. (1969) "Populations of Interacting Oscillators and Circadian Rhythms," *J. theoret. Biol., 22,* 418–436.

Pittendrigh, C.S. (1954) "On Temperature Independence in the Clock System Controlling Emergence Time in *Drosophila,*" *Proc. natn. Acad. Sci., U.S.A. 40,* 1018–1029.

Pittendrigh, C.S. (1960) "Circadian Rhythms and the Circadian Organization of Living Systems," *Cold Spring Harb. Symp. quant. Biol., 25,* 159–184.

Pittendrigh, C.S. (1966) "The Circadian Oscillation in *Drosophila pseudoobscura* Pupae: A Model for the Photoperiodic Clock," *Z. Pflanzenphysiol., 54,* 275–307.

Pittendrigh, C.S. and Minis, D.H. (1972) "Circadian Systems: Longevity as a function of Circadian Resonance in *Drosophila melanogaster,*" *Proc. natn. Acad. Sci., U.S.A., 69,* 1537–1539.

Pye, K. and Chance, B. (1966) "Sustained Sinusoidal Oscillations of Reduced Pyridine Nucleotide in a Cell-free Extract of *Saccharomyces carlsbergensis,*" *Proc. natn. Acad. Sci., U.S.A., 55,* 888–894.

Rensing, L. (1969) "Circadian Rhythms in Salivary Glands of *Drosophila, in vivo, in vitro,* and after addition or Ecdyson" (in German), *J. Insect Physiology, 15,* 2285–2303.

Saunders, D.S. (1972) "Circadian Control of Larval Growth Rate in *Sarcophaga argyrostoma,*" *Proc. natn. Acad. Sci., U.S.A., 69,* 2738–2740.

Sweeney, B.M. (1969) *Rhythmic Phenomena in Plants,* Academic Press, New York and London.

Turcote, D.L. Nordman, J.C. and Cisne, J.L. (1974) "Evolution of the Moon's Orbit and the Origin of Life," *Nature,* Lond., **251**, 124–125.

Tychsen, P.H. and Fletcher, B.S. (1971) "Studies on the Rhythm of Mating in the Queensland Fruit Fly *Dacus tryoni,*" *J. Insect Physiol.,* **17**, 2139–2156.

Wells, J.W. (1963) "Coral Growth and Geochronometry," *Nature,* Lond., **197**, 948–950.

Went, F.W. (1959) "The Periodic Aspect of Photoperiodism and Thermoperiodicity," in *Photoperiodism and Related Phenomena in Plants and Animals,* ed. Withrow, R.B., pp. 551-564, Am. Assn. Adv. Sci., Washington.

Wilkins, M.B. (1960) "The Effect of Light upon Plant Rhythms," *Cold Spring Harb. Symp. quant. Biol.,* **25**, 115–129.

Winfree, A.T. (1975) "On the Unclocklike Behaviour of Biological Clocks," *Nature,* Lond., **253**, 315–319.

Chapter 2

Aschoff, J. (1969) "Desynchronization and Resynchronization of Human Circadian Rhythms," *Aerosp. Med.,* **40**, 844–849.

Bateman, M.A. (1955) "The Effect of Light and Temperature on the Rhythm of Pupal Ecdysis in the Queensland Fruit-fly, *Dacus (Strumeta) tryoni* (Frogg)", *Austr. J. Zool.,* **3**, 22-33.

Bruce, V.G. (1972) "Mutants of the Biological Clock in *Chlamydomonas reinhardi,*" *Genetics,* **70**, 537–548.

Bünning, E. (1973) *The Physiological Clock,* Third English edition, Springer Verlag.

De Coursey, P.J. (1960) "Phase Control of Activity in a Rodent," *Cold Spring Harb. Symp. quant. Biol.,* **25**, 49–55.

Eskin, A. (1971) "Some Properties of the System Controlling the Circadian Activity Rhythm of Sparrows," in *Biochronometry,* ed. Menaker, M., pp. 55–80, National Academy of Sciences, Washington.

Gwinner, E. (1974) "Testosterone Induces "Splitting" of Circadian Locomotor Activity Rhythms in Birds," *Science,* Wash., **185**, 72–74.

Hoffmann, K. (1971) "Splitting of the Circadian Rhythm as a Function of Light Intensity," in *Biochronometry,* ed. Menaker, M., pp. 134–151, National Academy of Sciences, Washington.

Konopka, R. and Benzer, S. (1971) "Clock Mutants of *Drosophila melanogaster,*" *Proc. natn. Acad. Sci.,* U.S.A., **68**, 2112–2116.

Menaker, M. (1959) "Endogenous Rhythms of Body Temperature in Hibernating Bats," *Nature,* Lond., **184**, 1251–1252.

Minis, D.H. and Pittendrigh, C.S. (1968) "Circadian Oscillation Controlling Hatching: Its Ontogeny during Embryogenesis of a Moth," *Science,* Wash., **159**, 534–536.

Pittendrigh, C.S. (1954) "On Temperature Independence in the Clock System Controlling Emergence Time in *Drosophila,*" *Proc. natn. Acad. Sci.,* **40**, 1018–1029.

Pittendrigh, C.S. (1960) "Circadian Rhythms and the Circadian Organization of Living Systems," *Cold Spring Harb. Symp. quant. Biol.,* **25**, 159–184.

Pittendrigh, C.S. (1966) "The Circadian Oscillation in *Drosophila pseudoobscura* Pupae: A Model for the Photoperiodic Clock," *Z. Pflanzenphysiol.,* **54**, 275–307.

Pittendrigh, C.S. and Daan, S. (1974) "Circadian Oscillations in Rodents: a Systematic Increase of their Frequency with Age," *Science,* Wash., **186**, 548–550.

Rawson, K.S. (1960) "Effects of Tissue Temperature on Mammalian Activity Rhythms," *Cold Spring Harb. Symp. quant. Biol.,* **25**, 105–113.

Sargent, M.L. and Briggs, W. (1967) "The Effect of Light on the Circadian Rhythm of Conidiation in *Neurospora,*" *Plant Physiol.,* **42**, 1504–1510.

Skopik, S.D. and Pittendrigh, C.S. (1967) "Circadian Systems II. The Oscillation in the Individual *Drosophila* Pupa; its Independence of Developmental Stage," *Proc. natn. Acad. Sci.,* U.S.A., **58**, 1862–1869.

Taylor, B. and Jones, M.D.R. (1969) "The Circadian Rhythm of Flight Activity in the Mosquito *Aëdes aegypti* (L.): the Phase-setting Effect of Light-on and Light-off," *J. exp. Biol.,* **51,** 59–70.

Wilkins, M.B. (1960) "The Effect of Light on Plant Rhythms," *Cold Spring. Harb. Symp. quant. Biol.,* **25,** 115–129.

Winfree, A.T. (1974) "Suppressing *Drosophila*'s Circadian Rhythm with Dim Light," *Science,* Wash., **183,** 970–972.

Zimmerman, W.F. (1969) "On the Absence of Circadian Rhythmicity in *Drosophila pseudoobscura* Pupae," *Biol. Bull., Woods Hole,* **136,** 494–500.

Chapter 3

Aschoff, J. (1960) "Exogenous and Endogenous Components in Circadian Rhythms," *Cold Spring Harb. Symp. quant. Biol.,* **25,** 11–28.

Aschoff, J. (1965) "Response Curves in Circadian Periodicity," in *Circadian Clocks,* ed. Aschoff, J., pp. 95–111, North-Holland, Amsterdam.

Bruce, V.G. (1960) "Environmental Entrainment of Circadian Rhythms," *Cold Spring Harb. Symp. quant. Biol.,* **25,** 29–48.

Bruce, V.G. and Minis, D.H. (1969) "Circadian Clock Action Spectrum in a Photoperiodic Moth," *Science,* Wash., **163,** 583–585.

Chandrashekeran, M.K. (1967) "Studies on Phase Shifts in Endogenous Rhythms. I. Effects of Light Pulses on the Eclosion Rhythm," *Z. vergl. Physiol.,* **56,** 154–162.

Enright, J.T. (1965) "Synchronization and Ranges of Entrainment," in *Circadian clocks,* ed. Aschoff, J., pp. 112–124, North-Holland, Amsterdam.

Eskin, A. (1971) "Some Properties of the System Controlling the Circadian Activity Rhythm of Sparrows," in *Biochronometry,* ed. Menaker, M., pp. 55–80, National Academy of Sciences, Washington.

Frank, K.D. and Zimmerman, W.F. (1969) "Action Spectra for Phase Shifts of a Circadian Rhythm in *Drosophila*," *Science,* Wash., **163,** 688–689.

Lohmann, M. (1967) "Ranges of Circadian Period Length," *Experientia,* **23,** 788–790.

Maier, R. (1974) "Phase-shifting of the Circadian Rhythm of Eclosion in *Drosophila pseudoobscura* with Temperature Pulses," *J. interdiscipl. Cycle Res.,* **4,** 125–135.

Minis, D.H. (1965) "Parallel Peculiarities in the Entrainment of a Circadian Rhythm and Photoperiodic Induction in the Pink Bollworm (*Pectinophora gossypiella*)," in *Circadian Clocks,* ed. Aschoff, J., pp. 333–343, North-Holland, Amsterdam.

Pittendrigh, C.S. (1960) "Circadian Rhythms and the Circadian Organization of Living Systems," *Cold Spring Harb. Symp. quant. Biol.,* **25,** 159–184.

Pittendrigh, C.S. (1965) "On the Mechanism of Entrainment of a Circadian Rhythm by Light Cycles," in *Circadian Clocks,* ed. Aschoff, J., pp. 277–297, North-Holland, Amsterdam.

Pittendrigh, C.S. (1966) "The Circadian Oscillation in *Drosophila pseudoobscura* Pupae: A Model for the Photoperiodic Clock," *Z. Pflanzenphysiol.,* **54,** 275–307.

Pittendrigh, C.S. (1967) "Circadian Systems I. The Driving Oscillation and its Assay in *Drosophila pseudoobscura,*" *Proc. natn. Acad. Sci.,* U.S.A., **58,** 1762–1767.

Pittendrigh, C.S. and Bruce, V.G. (1959) "Daily Rhythms as Coupled Oscillator Systems and their relation to Thermoperiodism and Photoperiodism," in *Photoperiodism and Related Phenomena in Plants and Animals,* pp. 475–505, ed. Withrow, R.B., Amer. Assn. Adv. Sci., Washington.

Pittendrigh, C.S. and Minis, D.H. (1964) "The Entrainment of Circadian Oscillations by Light and their Role as Photoperiodic Clocks," *Amer. Nat.,* **98,** 261–294.

Roberts, S.K. de F. (1962) "Circadian Activity in Cockroaches II. Entrainment and Phase-shifting," *J. cell. Comp. Physiol.,* **59,** 175–186.

Sargent, M.L. and Briggs, W. (1967) "The Effect of Light on the Circadian Rhythm of Conidiation in *Neurospora,*" *Plant Physiol.,* **42,** 1504–1510.

Sweeney, B.M. and Hastings, J.W. (1960) "Effects of Temperature upon Diurnal Rhythms," *Cold Spring Harb. Symp. quant. Biol.,* **25**, 87–104.

Truman, J.W. (1971) "Circadian Rhythms and Physiology with special reference to Neuroendocrine Processes in Insects," *Proc. int. Symp. circadian Rhythmicity* (Wageningen), 11–135.

Winfree, A.T. (1971) "Corkscrews and Singularities in Fruitflies: Resetting Behavior of the Circadian Eclosion Rhythm," in *Biochronometry,* ed. Menaker, M., pp. 81–109, National Academy of Sciences, Washington.

Winfree, A.T. (1972) "Slow Dark-adaptation in *Drosophila*'s Circadian Clock," *J. comp. Physiol.,* **77**, 418–434.

Zimmerman, W.F. and Goldsmith, T.H. (1971) "Photosensitivity of the Circadian Rhythm and of Visual Receptors in Carotenoid Depleted *Drosophila*," *Science,* Wash., **171**, 1167–1168.

Zimmerman, W.F., Pittendrigh, C.S. and Pavlidis, T. (1968) "Temperature Compensation of the Circadian Oscillation in *Drosophila pseudoobscura* and its Entrainment by Temperature Cycles," *J. Insect Physiol.,* **14**, 669–684.

Chapter 4

Beier, W. and Lindauer, M. (1970) "The Sun's Position as a Zeitgeber for Bees" (in German), *Apidologie,* **1**, 5–28.

Beling, I. (1929) "On the Time-memory of Bees" (in German) *Z. vergl. Physiol.,* **9**, 259–338.

Bennett, M.F. and Renner, M. (1963) "The Collecting Performance of Honey Bees under Laboratory Conditions," *Biol. Bull., Woods Hole,* **125**, 416–430.

von Frisch, K. (1950) "The Sun as a Compass in the Lives of Bees" (in German), *Experientia,* **6**, 210–221.

von Frisch, K. and Lindauer, M. (1954) "Heaven and Earth in Competition in the Orientation of Bees" (in German), *Naturwiss.,* **41**, 245–253.

Hoffmann, K. (1960) "Experimental Manipulation of the Orientational Clock in Birds," *Cold Spring Harb. Symp. quant. Biol.,* **25**, 379–387.

Kleber, E. (1935) "Has the Time-memory of Bees a Biological Significance?" (in German). *Z. vergl. Physiol.,* **22**, 221–262.

Kramer, G. (1950) "Further Analysis of the Factors which Orientate Migration Activity of Caged Birds" (in German), *Naturwiss.,* **37**, 377–378.

Medugorac, I. and Lidauer, M. (1967) "The Time-memory of Bees under the influence of Narcosis and Social Zeitgebers" (in German), *Z. vergl. Physiol.,* **55**, 450–474.

Papi, F. (1960) "Orientation by Night: the Moon," *Cold Spring Harb. Symp. quant. Biol.,* **25**, 475–480.

Pardi, F. and Grassi, L. (1955) "Experimental Modification of Direction-finding in *Talitrus saltator* (Montague) and *Talorchestia deshayesei* (Aud.) (Crustacea-Amphipoda)," *Experientia,* **11**, 202.

Renner, M. (1955) "A Transoceanic Investigation of the Time-memory of the Honey-bee" (in German), *Naturwiss.,* **42**, 540–541.

Sauer, E.G.F. and Sauer, E.M. (1960) "Star Navigation of Nocturnal Migrating Birds." *Cold Spring Harb. Symp. quant. Biol.,* **25**, 463–473.

Chapter 5

Adkisson, P.L. (1964) "Action of the Photoperiod in Controlling Insect Diapause," *Amer. Nat.,* **98**, 357–374.

Danilevskii, A.S. (1965) *Photoperiodism and Seasonal Development of Insects,* First English Edition, Oliver and Boyd, Edinburgh and London.

Goryshin, N.I. and Tyshchenko, V.P. (1970) "Thermostability of the Process of Perception of Photoperiodic Information in the Moth *Acronycta rumicis* (Lepidoptera, Noctuidae)," *Dokl. Akad. Nauk SSSR,* **193**, 458–461 (in Russian).

Hamner, W.M. (1963) "Diurnal Rhythms and Photoperiodism in Testicular Recrudescence of the House Finch," *Science*, Wash., **142**, 1294–1294.

Hamner, W.M. (1964) "Circadian Control of Photoperiodism in the House Finch demonstrated by Interrupted-night Experiments," *Nature*, Lond., **203**, 1400–1401.

Lees, A.D. (1968) "Photoperiodism in Insects," in *Photophysiology*, Vol. IV, ed. Giese, A.C., pp. 47-137, Academic Press, New York.

Lees, A.D. (1971) "The Relevance of Action Spectra in the Study of Insect Photoperiodism," in *Biochronometry*, ed. Menaker, M. pp. 372–380, National Academy of Sciences, Washington.

Lees, A.D. (1973) "Photoperiodic Time Measurement in the Aphid *Megoura viciae*," *J. Insect Physiol.*, **19**, 2279–2316.

Menaker, M. and Eskin, A. (1967) "Circadian Clock in Photoperiodic Time Measurement: a Test of the Bünning Hypothesis," *Science*, Wash., **157**, 1182–1185.

Pittendrigh, C.S. (1960) "Circadian Rhythms and the Circadian Organization of Living Systems," *Cold Spring Harb. Symp. quant. Biol.*, **25**, 159–184.

Pittendrigh, C.S. (1966) "The Circadian Oscillation in *Drosophila pseudoobscura* Pupae: A Model for the Photoperiodic Clock," *Z. Pflanzenphysiol.*, **54**, 275–307.

Pittendrigh, C.S. (1972) "Circadian Surfaces and the Diversity of Possible Roles of Circadian Organization in Photoperiodic Induction," *Proc. natn. Acad. Sci.*, U.S.A., **69**, 2734–2737.

Pittendrigh, C.S. and Minis, D.H. (1964) "The Entrainment of Circadian Oscillations by Light and their Role as Photoperiodic Clocks," *Amer. Nat.*, **98**, 261–294.

Pittendrigh, C.S. and Minis, D.H. (1971) "The Photoperiodic Time Measurement in *Pectinophora gossypiella* and its relation to the Circadian System in that Species," in *Biochronometry*, ed. Menaker, M. pp. 212–250, National Academy of Sciences, Washington.

Saunders, D.S. (1970) "Circadian Clock in Insect Photoperiodism," *Science*, Wash., **168**, 601–603.

Saunders, D.S. (1971) "The Temperature-compensated Photoperiodic Clock 'Programming' Development and Pupal Diapause in the Flesh Fly *Sarcophaga argyrostoma*," *J. Insect Physiol.*, **17**, 801–812.

Saunders, D.S. (1974) "Circadian Rhythms and Photoperiodism in Insects," in *The Physiology of Insecta*, ed. Rockstein, M., pp. 461–533, Academic Press, New York.

Saunders, D.S. (1975*a*) "'Skeleton' Photoperiods and the Control of Diapause and Development in the Flesh Fly, *Sarcophaga argyrostoma*," *J. comp. Physiol.*, **97**, 97–112.

Saunders, D.S. (1975*b*) "Spectral Sensitivity and Intensity Thresholds in *Nasonia* Photoperiodic Clock," *Nature*, Lond., **253**, 732–734.

Zdarek, J. and Slama, K. (1972) "Supernumerary Larval Instars in Cyclorrhaphous Diptera," *Biol. Bull., Woods Hole*, **142**, 350–357.

Chapter 6

Blake, G.M. (1959) "Control of Diapause by an 'Internal Clock' in *Anthrenus verbasci* (L.) (Col., Dermestidae)," *Nature*, Lond., **183**, 126–127.

Goss, R.J. (1969) "Photoperiodic Control of Antler Cycles in Deer", (i) *J. exp. Zool.*, **170**, 311–324, (ii) *J. exp. Zool.*, **171**, 233–234.

Gwinner, E. (1971) "A Comparative Study of Circannual Rhythms in Warblers," in *Biochronometry*, ed. Menaker, M., pp. 405–427, National Academy of Sciences, Washington.

Heller, H.C. and Poulson, T.L. (1970) "Circannian Rhythms—II. Endogenous and Exogenous Factors controlling Reproduction and Hibernation in Chipmunks (*Eutamias*) and Ground Squirrels (*Spermophilus*), "*Comp. Biochem. Physiol.*, **33**, 357–383.

Jegla, T.C. and Poulson, T.L. (1970) "Circannian Rhythms—I. Reproduction in the Cave Crayfish, *Orconectes pellucidus inermis*," *Comp. Biochem. Physiol.*, **33**, 347–355.

Lofts, B. (1964) "Evidence of an Autonomous Reproductive Rhythm in an Equatorial Bird (*Quelea quelea*)," *Nature*, Lond., **201**, 523–524.

Pengelley, E.T. and Asmundson, S.M. (1969) "Free-running Periods of Endogenous Circannian Rhythms in the Golden-mantled Ground Squirrel, *Citellus lateralis*," *Comp. Biochem. Physiol.*, **30**, 177–183.

Pengelley, E.T. and Kelly, K.H. (1966) "A 'Circannian' Rhythm in Hibernating Species of the Genus *Citellus* with observations on their Physiological Evolution," *Comp. Biochem. Physiol.*, **19**, 603–617.

Schwab, R.G. (1971) "Circannian Testicular Periodicity in the European Starling in the absence of Photoperiodic Change," in *Biochronometry*, ed. Menaker, M., pp. 428–447, National Academy of Sciences, Washington.

Chapter 7

Bünning, E. and Müller, D. (1961) "How do Organisms Measure Lunar Cycles?" (in German), *Z. Naturforsch.*, **16b**, 391–395.

Enright, J.T. (1961) "Pressure Sensitivity of an Amphipod," *Science*, Wash., **133**, 758–760.

Enright, J.T. (1963) "The Tidal Rhythm of Activity of a Sand-beach Amphipod," *Z. vergl. Physiol.*, **46**, 276–313.

Enright, J.T. (1965) "Entrainment of A Tidal Rhythm," *Science*, Wash., **147**, 864–867.

Enright, J.T. (1972) "A Virtuoso Isopod: Circa-lunar Rhythms and their Tidal Fine Structure," *J. comp. Physiol.*, **77**, 141–162.

Fingerman, M. (1960) "Tidal Rhythmicity in Marine Organisms," *Cold Spring Harb. Symp. quant. Biol.*, **25**, 481–489.

Gibson, R.N. (1965) "Rhythmic Activity in Littoral Fish," *Nature*, Lond., **207**, 544–545.

Hartland-Rowe, R. (1955) "Lunar Rhythm in the Emergence of an Ephemeropteran," *Nature*, Lond., **176**, 657.

Hauenschild, C. (1960) "Lunar Periodicity," *Cold Spring Harb. Symp. quant. Biol.*, **25**, 491–497.

Klapow, L.A. (1972) "Natural and Artificial Rephasing of a Tidal Rhythm," *J. comp. Physiol.*, **79**, 233–258.

Naylor, E. (1958) "Tidal and Diurnal Rhythms of Locomotor Activity in *Carcinus maenas* (L)," *J. exp. Biol.*, **35**, 602–610.

Naylor, E. (1963) "Temperature Relationships of the Locomotor Rhythm of *Carcinus*," *J. exp. Biol.*, **40**, 669–679.

Naylor, E. and Atkinson, R.J.A. (1972) "Pressure and the Rhythmic Behaviour of Inshore Marine Animals," *Symp. Soc. exp. Biol.*, **26**, 395–415.

Neumann, D. (1966) "The Lunar and Daily Hatching Period of the Midge *Clunio*. Control and Timing of the Tidal Periodicity" (in German), *Z. vergl. Physiol.*, **53**, 1–61.

Palmer, J.D. (1973) "Tidal Rhythms: the Clock Control of the Physiology of Marine Organisms," *Biol. Rev.*, **48**, 377–418.

Palmer, J.D. and Round, F.E. (1967) "Persistent, Vertical Migration Rhythms in Benthic Microflora VI. The Tidal and Diurnal Nature of the Rhythm in the Diatom *Hantzschia virgata*," *Biol. Bull.*, Woods Hole, **132**, 44–55.

Rao, K.P. (1954) "Tidal Rhythmicity of Rate of Water Propulsion in *Mytilus*, and its Modifiability by Transplantation," *Biol. Bull.*, Woods Hole, **106**, 353–359.

Youthed, G.J. and Moran, V.C. (1969) "The Lunar-day Activity Rhythm of Myrmeleontid Larvae," *J. Insect Physiol.*, **15**, 1259–1271.

Chapter 8

Brady, J.N. (1969) "How are Insect Circadian Rhythms Controlled?" *Nature*, Lond., **223**, 781–784.

Brady, J.N. (1974) "The Physiology of Insect Circadian Rhythms," *Adv. Insect Physiol.*, **10**, 1–115.

Cymborowski, B. and Brady, J.N. (1973) "Insect Circadian Rhythms transmitted by Parabiosis—a Re-examination," *Nature, New Biology*, **236**, 221–222.

Gaston, S. (1971) "The Influence of the Pineal Organ on the Circadian Activity Rhythm in Birds," in *Biochronometry*, ed. Menaker, M., pp. 541–548, National Academy of Sciences, Washington.

Harker, J.E. (1960) "Endocrine and Nervous Factors in Insect Circadian Rhythms," *Cold Spring Harb. Symp. quant. Biol.*, **25**, 279–287.

Jacklet, J.W. (1969) "Circadian Rhythm of Optic Nerve Impulses Recorded in Darkness from Isolated Eye of *Aplysia*," *Science, Wash.*, **164**, 562–563.

Lees, A.D. (1964) "The Location of the Photoperiodic Receptors in the Aphid *Megoura viciae* Buckton," *J. exp. Biol.*, **41**, 119–133.

Lickey, M.E., Zack, S. and Birrell, P. (1971) "Some Factors Governing Entrainment of a Circadian Rhythm in a Single Neuron," in *Biochronometry*, ed. Menaker, M., pp. 549–564, National Academy of Sciences, Washington.

Loher, W. (1972) "Circadian Control of Stridulation in the Cricket *Teleogryllus commodus* Walker," *J. comp. Physiol.*, **79**, 173–190.

Menaker, M. (1971) "Synchronization with the Photic Environment via Extraretinal Receptors in the Avian Brain," in *Biochronometry*, ed. Menaker, M., pp. 315–332, National Academy of Sciences, Washington.

Menaker, M. and Keatts, H. (1968) "Extraretinal Light Perception in the Sparrow II. Photoperiodic Stimulation of Testis Growth," *Proc. natn. Acad. Sci.*, U.S.A., **60**, 146–151.

Moore, R.Y., Heller, A., Wurtman, R.J. and Axelrod, J. (1967) "Visual Pathway Mediating Pineal Response to Environmental Light," *Science, Wash.*, **155**, 220–223.

Naylor, E. and Williams, B.G. (1968) "Effects of Eyestalk Removal on Rhythmic Locomotor Activity in *Carcinus*," *J. exp. Biol.*, **49**, 107–116.

Nishiitsutsuji-Uwo, J. and Pittendrigh, C.S. (1968) "Central Nervous System Control of Circadian Rhythmicity in the Cockroach," II., *Z. vergl. Physiol.*, **58**, 1–13; III, *Z. vergl. Physiol.*, **58**, 14–46.

Roberts, S.K. de F. (1965) "Photoreception and Entrainment of Cockroach Activity Rhythms," *Science, Wash.*, **148**, 958–959.

Roberts, S.K. de F. (1974) "Circadian Rhythms in Cockroaches. Effects of Optic Lobe Lesions," *J. comp. Physiol.*, **88**, 21–30.

Strumwasser, F. (1965) "The Demonstration and Manipulation of a Circadian Rhythm in a Single Neuron," in *Circadian Clocks*, ed. Aschoff, J., pp. 442–462, North-Holland, Amsterdam.

Truman, J.W. (1971) "Circadian Rhythms and Physiology with special reference to Neuroendocrine Processes in Insects," *Proc. int. Symp. circadian Rhythmicity* (Wageningen), 111–135.

Truman, J.W. and Riddiford, L.M. (1970) "Neuroendocrine Control of Ecdysis in Silkmoths," *Science, Wash.*, **167**, 1624–1626.

Williams, C.M. and Adkisson, P.L. (1964) "Physiology of Insect Diapause–XIV. An Endocrine Mechanism for the Photoperiodic Control of Pupal Diapause in the Oak Silkworm, *Antheraea pernyi*," *Biol. Bull, Woods Hole*, **127**, 511–525.

Zimmerman, W.F. and Ives, D. (1971) "Some Photophysiological Aspects of Circadian Rhythmicity in *Drosophila*," in *Biochronometry*, ed. Menaker, M., pp. 381–391, National Academy of Sciences, Washington.

Chapter 9

Brenner, W. and Engelmann, W. (1973) "Heavy Water Slows Down the Photoperiodic Timing of Flower Induction in *Chenopodium rubrum*, " *Z. Naturforsch.*, **28c**, 356.

Brinkmann, K. (1971) "Metabolic Control of Temperature Compensation in the Circadian Rhythm of *Euglena gracilis*," in *Biochronometry*, ed. Menaker, M., pp. 567–593, National Academy of Sciences, Washington.

Bruce, V.G. and Pittendrigh, C.S. (1960) "An Effect of Heavy Water on the Phase and Period of the Circadian Rhythm in *Euglena*," *J. cell. comp. Physiol.*, **56**, 25–31.

Bühnemann, F. (1955) "Das endodiurnale System der Oedogoniumzelle—II. Der Einfluss von Stoffwechselgiften und anderen Wirkstoffen," *Biol. Zentralblatt.*, **74**, 691–705.

Ehret, C.F. and Trucco, E. (1967) "Molecular Models for the Circadian Clock," *J. theoret. Biol.*, **15**, 240–262.

Engelmann, W. (1972) "Lithium Slows Down the *Kalanchoë Clock*," *Z. Naturforsch.*, **27**b, 477.

Enright, J.T. (1971*a*) "Heavy Water Slows Down Biological Timing Process," *Z. vergl. Physiol.*, **72**, 1–16.

Enright, J.T. (1971*b*) "The Internal Clock of Drunken Isopods," *Z. vergl. Physiol.*, **75**, 332–346.

Feldman, J. (1967) "Lengthening the Period of a Biological Clock in *Euglena* by Cyclohexi-mide, an Inhibitor of Protein Synthesis," *Proc. natn. Acad. Sci.*, U.S.A., **57**, 1080–1087.

Hastings, J.W. (1970) "Cellular-biochemical Clock Hypothesis," in *The Biological Clock: Two Views*, ed. Palmer, J.D., pp. 61–91, Academic Press, New York and London.

Hastings, J.W. and Keynan, A. (1965) "Molecular Aspects of Circadian Systems," in *Circadian Clocks*, ed. Aschoff, J., pp. 167–182, North-Holland, Amsterdam.

Jacklet, J.W. and Geronimo, J. (1971) "Circadian Rhythms: Populations of Interacting Neurons," *Science*, Wash., **174**, 299–302.

Neville, A.C. (1970) "Cuticle Ultrastructure in relation to the Whole Insect," *Symp. R. ent. Soc. Lond.*, **5**, 17–39.

Njus, D., Sulzman, F.M. and Hastings, J.W. (1974) "Membrane Model for the Circadian Clock," *Nature*, Lond., **248**, 116–120.

Pavlidis, T. (1969) "Populations of Interacting Oscillators and Circadian Rhythms," *J. theoret. Biol.*, **22**, 418–436.

Pittendrigh, C.S., Caldarola, P.C. and Cosbey, E.S. (1973) "A Differential Effect of Heavy Water on Temperature-dependent and Temperature-compensated Aspects of the Circadian System of *Drosophila pseudoobscura*," *Proc. natn. Acad. Sci.*, U.S.A., **70**, 2037–2041.

Pye, K. and Chance, B. (1966) "Sustained Sinusoidal Oscillations of Reduced Pyridine Nucleotide in a Cell-free Extract of *Saccharomyces carlsbergensis*," *Proc. natn. Acad. Sci.*, U.S.A., **55**, 888–894.

Schweiger, H.G. (1971) "Circadian Rhythms: Subcellular and Biochemical Aspects," *Proc. int. Symp. circadian Rhythmicity* (Wageningen), 157–174.

Suter, R.B. and Rawson, K.S. (1968) "Circadian Activity Rhythm of the Deer Mouse, *Peromyscus*: Effect of Deuterium Oxide," *Science*, Wash., **160**, 1011–1014.

Vanden Driessche, T. (1971) "Structural and Functional Rhythms in the Chloroplasts of *Acetabularia*: Molecular Aspects of the Circadian System," in *Biochronometry*, ed. Menaker M., pp. 612–622, National Academy of Sciences, Washington.

Winfree, A.T. (1967) "Biological Rhythms and the Behaviour of Populations of Coupled Oscillators," *J. theoret. Biol.*, **16**, 15–42.

Winfree, A.T. (1975) "On the Unclocklike Behaviour of Biological Clocks," *Nature*, Lond., **253**, 315–319.

GLOSSARY

Aschoff's rule. The period of a free-running biological oscillation (τ) lengthens on transfer from DD to *LL* or with an increase in light intensity for dark-active animals, but shortens for light-active animals.

Circadian (rhythm). An endogenous biological oscillation with a natural period (τ) *close to,* but not necessarily equal to, that of the solar day (24 hours).

Circadian time (Ct). Time scale (in hours, or subdivisions of the circadian period) covering one full period of an oscillation (i.e. Ct 01 to Ct 24).

Circatidal, circasyzygic, circalunar, circannual rhythms. Endogenous biological oscillations with a natural period (τ) close to the tidal cycle (12.4 hours), the spring tide to spring tide cycle (14.7 days), the lunar cycle (29 days), or the year, respectively.

Diapause. A period of arrest of growth and development which enables insects to overwinter (hibernate) or withstand drought (aestivate), or to synchronize their developmental cycles with that of the seasons. Diapause involves the cessation of neuroendocrine activity, and is most frequently induced by photoperiod.

Entrainment. In the context of this book, entrainment is the synchronization of a biological oscillation to a Zeitgeber so that both have the same period.

External coincidence. A model for the photoperiodic clock in which light has a dual role: (1) it entrains and hence phase-sets the photoperiodic oscillations, and (2) it controls photoperiodic induction by a temporal coincidence with a photoperiodically-inducible phase (ϕ_i).

Free-running period (τ). The period of an endogenous oscillator revealed in the absence of a Zeitgeber (e.g. at constant temperature, and in continuous darkness (DD) or continuous light (*LL*)).

Gate. The "allowed zone" of the cycle, dictated by the circadian clock, through which flies may emerge, hatch, etc.

Hour-glass (interval timer). A non-repetitive (i.e. non-oscillatory) timer which is set in motion at, say, dawn or dusk, and then runs its course, rather like an egg timer.

Internal coincidence. A model for the photoperiodic clock in which two or more circadian oscillators are independently phase-set by dawn and dusk, and photoperiodic induction depends on the phase angle between them.

Period. The time after which a definite phase reoccurs (see figure G.1a). In biological systems it should be stated which overt phase is being used to determine period (e.g. onset of activity, median of eclosion peaks, etc.).

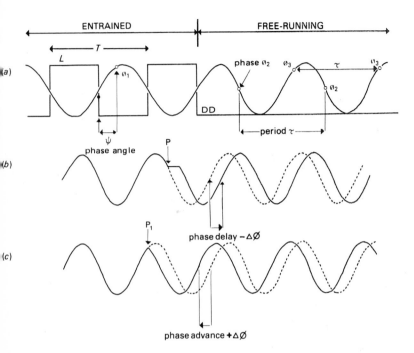

Figure G.1a A biological oscillation (sine wave) firstly entrained by an environmental Zeitgeber (square wave), and then free-running in constant darkness.
D = dark; L = light; DD = constant darkness; T = period of Zeitgeber (in this case one "dawn" to the next, 24 hours).
● – a phase point of the Zeitgeber oscillation ("dusk").
o– different phase points of the biological oscillation (ϕ_1, ϕ_2 etc).
ø = phase; τ = period of oscillation in free-run measured from one phase (ϕ_1) to its next occurrence. Note that in this illustration τ in free-run is less than 24 hours.
ψ = phase angle, in this illustration the phase angle between dusk and ϕ_1.
(b) A biological oscillation in free-run subjected to a perturbation P causing a delay phase-shift ($-\Delta\phi$).
(c) A biological oscillation in free-run subjected to a perturbation P_1 causing an advance phase-shift ($+\Delta\phi$).

Phase (ø). The instantaneous state of an oscillation within a period (figure G.1a). In a circadian rhythm it may be the onset of locomotor activity, the peak of pupal eclosion, the point of sensitivity to light, etc.

Phase angle (ψ). The phase relationship between two phases on the same or different oscillations. In figure G.1a, for example, the phase angle between a phase of the Zeitgeber cycle ("dusk") and a phase of the biological oscillation (ø₁) is shown as ψ, and normally measured in hours or in fractions of the circadian period.

Phase response curve (PRC). A plot of phase-shift (magnitude and sign) caused by a single perturbation (i.e. light or temperature pulses) at different circadian phases (circadian times) of an oscillator in free-run.

Phase-shift (Δø). A single displacement of an oscillation along the time-axis following a perturbation. It may involve either an advance (+Δø) or a delay (−Δø) (figure G.1b and c).

Photoperiod. The period of light in the daily cycle (daylength), measured in hours.

Photoperiodic counter. That aspect of the photoperiodic response which consists of a temperature-compensated mechanism which accumulates "information" from successive photoperiodic cycles.

Photoperiodic response curve. The response of a population of an organism to a range of stationary photoperiods (DD to LL) usually including the critical daylength.

Photoperiodically inducible phase (øi). A hypothetical phase-point in an oscillator (or perhaps a driven rhythm) which is light-sensitive, and an integral part of the external-coincidence model.

Range of entrainment. Range of frequencies within which a biological oscillation can be entrained by a Zeitgeber. For most organisms the range (for circadian rhythms) is 18–30 hours.

Required day number (RDN). The temperature-compensated number of inductive photoperiods in the photoperiodic counter required to raise the incidence of diapause in a particular day's batch to 50%.

Skeleton photoperiod. A light regime using two shorter pulses of light to simulate dawn and dusk effects of a longer complete photo-period. Skeleton photoperiods may be either symmetrical (i.e. composed of two pulses of equal duration) or asymmetrical.

Thermoperiod. A daily temperature cycle.

Transients. One or more temporarily shortened or lengthened periods following perturbation by a light or temperature pulse.

Zeitgeber (time-giver). The forcing oscillation which entrains a biological oscillation, e.g. the environmental cycles of light and temperature, tide, moonlight, and season.

Zeitgedächtnis. The "time-memory" of bees.

Zugunruhe. Restlessness of captive birds at the migration season.

Symbols

L	Light fraction of cycle
D	Dark fraction of cycle
LD	Light/dark cycle.
LD	*12*:12 Represents 12 hours of light and 12 hours of darkness in each 24-hour cycle.
LL	Continuous light
DD	Continuous darkness
τ	Natural period of a biological oscillator as revealed in "free-running" conditions.
T	Period of Zeitgeber (in the case of the day-night cycle, 24 hours).
Q	Phase point
Q_i	The photoperiodically inducible phase
ψ	Phase angle or phase relation
ΔQ	Phase-shift, + or −
$-\Delta\varnothing$	Delay phase-shift
$+\Delta\varnothing$	Advance phase-shift
α	Activity time
ρ	Rest time
θ	Cophase. The time-interval between the end of the light perturbation and the eclosion peaks in *Drosophila*.
Ct	Circadian time
Zt	Zeitgeber time.

Index